Communications in Computer and Information Science 2492

Series Editors

Gang Li, *School of Information Technology, Deakin University, Burwood, VIC, Australia*
Joaquim Filipe, *Polytechnic Institute of Setúbal, Setúbal, Portugal*
Zhiwei Xu, *Chinese Academy of Sciences, Beijing, China*

Rationale
The CCIS series is devoted to the publication of proceedings of computer science conferences. Its aim is to efficiently disseminate original research results in informatics in printed and electronic form. While the focus is on publication of peer-reviewed full papers presenting mature work, inclusion of reviewed short papers reporting on work in progress is welcome, too. Besides globally relevant meetings with internationally representative program committees guaranteeing a strict peer-reviewing and paper selection process, conferences run by societies or of high regional or national relevance are also considered for publication.

Topics
The topical scope of CCIS spans the entire spectrum of informatics ranging from foundational topics in the theory of computing to information and communications science and technology and a broad variety of interdisciplinary application fields.

Information for Volume Editors and Authors
Publication in CCIS is free of charge. No royalties are paid, however, we offer registered conference participants temporary free access to the online version of the conference proceedings on SpringerLink (http://link.springer.com) by means of an http referrer from the conference website and/or a number of complimentary printed copies, as specified in the official acceptance email of the event.

CCIS proceedings can be published in time for distribution at conferences or as post-proceedings, and delivered in the form of printed books and/or electronically as USBs and/or e-content licenses for accessing proceedings at SpringerLink. Furthermore, CCIS proceedings are included in the CCIS electronic book series hosted in the SpringerLink digital library at http://link.springer.com/bookseries/7899. Conferences publishing in CCIS are allowed to use Online Conference Service (OCS) for managing the whole proceedings lifecycle (from submission and reviewing to preparing for publication) free of charge.

Publication process
The language of publication is exclusively English. Authors publishing in CCIS have to sign the Springer CCIS copyright transfer form, however, they are free to use their material published in CCIS for substantially changed, more elaborate subsequent publications elsewhere. For the preparation of the camera-ready papers/files, authors have to strictly adhere to the Springer CCIS Authors' Instructions and are strongly encouraged to use the CCIS LaTeX style files or templates.

Abstracting/Indexing
CCIS is abstracted/indexed in DBLP, Google Scholar, EI-Compendex, Mathematical Reviews, SCImago, Scopus. CCIS volumes are also submitted for the inclusion in ISI Proceedings.

How to start
To start the evaluation of your proposal for inclusion in the CCIS series, please send an e-mail to ccis@springer.com.

Pierangela Bruno · Francesco Calimeri ·
Francesco Cauteruccio · Giorgio Terracina
Editors

Hybrid Models for Coupling Deductive and Inductive Reasoning

Third International Workshop, HYDRA 2024
Santiago de Compostela, Spain, October 20, 2024
Revised Selected Papers

Editors
Pierangela Bruno ⓘ
University of Calabria
Rende, Italy

Francesco Calimeri ⓘ
University of Calabria
Rende, Italy

Francesco Cauteruccio ⓘ
University of Salerno
Fisciano, Italy

Giorgio Terracina ⓘ
University of Calabria
Rende, Italy

ISSN 1865-0929   ISSN 1865-0937 (electronic)
Communications in Computer and Information Science
ISBN 978-3-031-89365-0   ISBN 978-3-031-89366-7 (eBook)
https://doi.org/10.1007/978-3-031-89366-7

© The Editor(s) (if applicable) and The Author(s), under exclusive license
to Springer Nature Switzerland AG 2025

This work is subject to copyright. All rights are solely and exclusively licensed by the Publisher, whether the whole or part of the material is concerned, specifically the rights of translation, reprinting, reuse of illustrations, recitation, broadcasting, reproduction on microfilms or in any other physical way, and transmission or information storage and retrieval, electronic adaptation, computer software, or by similar or dissimilar methodology now known or hereafter developed.
The use of general descriptive names, registered names, trademarks, service marks, etc. in this publication does not imply, even in the absence of a specific statement, that such names are exempt from the relevant protective laws and regulations and therefore free for general use.
The publisher, the authors and the editors are safe to assume that the advice and information in this book are believed to be true and accurate at the date of publication. Neither the publisher nor the authors or the editors give a warranty, expressed or implied, with respect to the material contained herein or for any errors or omissions that may have been made. The publisher remains neutral with regard to jurisdictional claims in published maps and institutional affiliations.

This Springer imprint is published by the registered company Springer Nature Switzerland AG
The registered company address is: Gewerbestrasse 11, 6330 Cham, Switzerland

If disposing of this product, please recycle the paper.

Preface

In the rapidly evolving domain of artificial intelligence (AI), the interplay between deductive and inductive reasoning methods presents a fascinating opportunity for exploration. Deductive reasoning, grounded in explicit premises and formal logical inference, contrasts with inductive reasoning, which derives generalizations from observations, often leveraging machine learning and deep learning techniques. The convergence of these paradigms holds the potential to unlock transformative capabilities for AI systems, enhancing their adaptability and resilience in addressing a wide range of complex challenges.

The International Workshop on Hybrid Models for Coupling Deductive and Inductive Reasoning (HYDRA) was designed as a forum for researchers to explore the exciting possibilities at the intersection of deductive and inductive reasoning.

The third edition of HYDRA marked a significant milestone in advancing the integration of these two core foundations of AI research. This collection of proceedings brings together a diverse array of research contributions, perspectives, and insights from the scientific community engaged in this area. Our workshop invited original contributions spanning theoretical models, practical applications, and empirical findings. We aimed to highlight particularly compelling approaches that address significant challenges, such as developing methods to seamlessly integrate logical and statistical frameworks, designing algorithms for reasoning under incomplete or uncertain information, and creating tools to interpret and explain hybrid models. Furthermore, we encouraged researchers to examine the ethical and societal dimensions of these emerging technologies, including fairness, accountability, and transparency. The workshop covered a variety of topics in its call for submissions:

- Hybrid inductive-deductive approaches to AI,
- Interaction of inductive and deductive techniques for AI solutions,
- Integration of Answer Set Programming (ASP) in inductive scenarios,
- Integration of Constraint Programming (CSP) in inductive scenarios,
- Integration of other logic programming paradigms in inductive scenarios,
- Integration of declarative solutions in inductive scenarios,
- Logic programming language extensions for supporting inductive processes,
- New methods for coupling peculiarities of deductive and inductive systems,
- Inductive reasoning to enhance and improve deductive systems,
- Deductive processes for intensive data flow management,
- Deductive processes in strong inductive-tailored scenarios,
- Knowledge representation and reasoning for improving and enhancing inductive processing,
- Discussions and positions on novel hybrid methods of deductive and inductive reasoning,
- Evaluation and comparison of existing deductive and inductive methods,
- Hybridizing logic programming paradigms with procedural approaches,

- Novel contexts of application for hybrid deductive and inductive systems,
- Neuro-symbolic approach to reasoning and learning.

The workshop papers underwent a single-blind peer-review process, with each submission assigned to three members of the Program Committee. Ultimately, six papers were chosen out of eight total submissions for oral presentation.

Many people and organizations contributed to the success of HYDRA 2024; we would like to extend our heartfelt gratitude to all of them. We sincerely thank the Chairs and the Organizers of the 27th European Conference on Artificial Intelligence (ECAI 2024) for hosting the workshop, and Springer for publishing these proceedings. We are also deeply grateful to the Program Committee members for their insightful and invaluable reviews, the dedicated authors for their active participation, and the enthusiastic attendees for their engagement. These collective contributions significantly enriched the workshop experience, advancing HYDRA as a reference point for scholars working on or interested in bridging the gap between deductive and inductive approaches to AI.

December 2024

Pierangela Bruno
Francesco Calimeri
Francesco Cauteruccio
Giorgio Terracina

Organization

General Chairs

Francesco Calimeri University of Calabria, Italy
Giorgio Terracina University of Calabria, Italy

Program Chairs

Pierangela Bruno University of Calabria, Italy
Francesco Cauteruccio University of Salerno, Italy

Publicity Chair

Weronika T. Adrian AGH University of Science and Technology, Poland

Technical Chair

Edoardo De Rose University of Calabria, Italy

Program Committee

Mario Alviano University of Calabria, Italy
Alessia Amelio University "G. d'Annunzio" Chieti-Pescara, Italy
Andreas Brännström Umeå University, Sweden
Luciano Caroprese University "G. d'Annunzio" Chieti-Pescara, Italy
Stefano Cirillo Universita degli Studi di Salerno, Italy
Richard Comploi-Taupe Siemens AG Österreich, Austria
Erica Coppolillo University of Calabria, Italy
Stefania Costantini University of L'Aquila, Italy
Carmine Dodaro University of Calabria, Italy
Mauro Dragoni Fondazione Bruno Kessler - FBK-IRST, Italy
Jonas Ehrhardt Helmut-Schmidt University, Germany
Esra Erdem Sabanci University, Turkey

Wolfgang Faber	Alpen-Adria-Universität Klagenfurt, Austria
Massimo Guarascio	ICAR-CNR, Italy
René Heesch	Helmut Schmidt University, Germany
Antonio Ielo	University of Calabria, Italy
Thomas Lukasiewicz	University of Oxford, UK
Marco Maratea	University of Calabria, Italy
Cinzia Marte	University of Calabria, Italy
Aldo Marzullo	Humanitas Research Hospital - AI Center, Italy
Giuseppe Mazzotta	University of Calabria, Italy
Oliver Niggemann	Helmut-Schmidt-Universität/Universität der Bundeswehr Hamburg, Germany
Luca Oneto	University of Genoa, Italy
Pierpaolo Pellicori	University of Glasgow, UK
Rafael Peñaloza	University of Milano-Bicocca, Italy
Luigi Portinale	Università del Piemonte Orientale "A. Avogadro", Italy
Francesco Ricca	University of Calabria, Italy
Sabrina Senatore	University of Salerno, Italy
Gioacchino Sterlicchio	Polytechnic University of Bari, Italy
Domenico Ursino	Polytechnic University of the Marches, Italy
Alessio Zanga	University of Milano - Bicocca, Italy
Ester Zumpano	University of Calabria, Italy

Contents

Answer Set Programming and Neurosymbolic AI: Applications and Future
Perspectives (Invited Talk) 1
   Manuel Borroto, Antonio Ielo, Giuseppe Mazzotta, and Francesco Ricca

Understanding Artificial Intelligence in Chess: The RubiChess Case Study 22
   Davide Tosi and Marika Scalise

Evaluating Inductive Reasoning Capabilities of Large Language Models
With The One Dimensional Abstract Reasoning Corpus 35
   Cédric Mesnage, Xiaoyang Wang, Hang Dong, and Aishwaryaprajna

Program Synthesis Using Inductive Logic Programming for the Abstraction
and Reasoning Corpus 52
   *Filipe Marinho Rocha, Inês Dutra, Vítor Santos Costa,
   and Luís Paulo Reis*

Trustworthy Inductive Knowledge for Tropical Cyclones Formation
Detection 69
   *Marco Adriano Ferrero, Davide Azzalini, Matteo Giuliani,
   Andrea Castelletti, Federico Cerutti, and Francesco Amigoni*

Automatic Curriculum Cohesion Analysis Based on Knowledge Graphs 82
   Paulina Gacek and Weronika T. Adrian

Online Inductive Learning from Answer Sets for Efficient Reinforcement
Learning Exploration 93
   Celeste Veronese, Daniele Meli, and Alessandro Farinelli

Author Index 107

Answer Set Programming and Neurosymbolic AI: Applications and Future Perspectives (Invited Talk)

Manuel Borroto, Antonio Ielo(✉), Giuseppe Mazzotta, and Francesco Ricca

University of Calabria, Arcavacata (CS), Cosenza, Italy
{manuel.borroto,antonio.ielo,giuseppe.mazzotta,francesco.ricca}@unical.it

Abstract. Answer Set Programming (ASP) is a well-known symbolic AI formalism developed in the area of knowledge representation and reasoning. This paper reports on some blendings of ASP with neural approaches, that can be classified as neurosymbolic AI systems. In particular, we describe (*i*) an industrial application of ASP and neural network that addresses the compliance-checking of electrical control panels, (*ii*) two possible combinations of ASP with large language models for enabling reasoning from text, and (*iii*) the usage of machine learning techniques to boost the evaluation of ASP programs by implementing solver selection pipelines.

Keywords: Answer Set Programming · neurosymbolic AI · Applications

1 Introduction

Answer Set Programming (ASP) [24] is a declarative AI formalism for knowledge representation and reasoning. ASP is based on solid theoretical foundations [8] and has been successfully applied to several academic and industrial AI applications such as planning, scheduling, robotics, decision support, natural language understanding, and more (cfr. [21,22,88]). Indeed, ASP provides an effective approach for representing and solving various classes of problems, leveraging a standardized first-order language implemented in several highly efficient systems [10]. As a matter of fact ASP is a symbolic AI formalism that is particularly useful for solving combinatorial optimization problems and knowledge-intensive applications.

In recent years, there has been a significant surge of interest in integrating Knowledge Representation techniques with Machine Learning methods [99]. This growing interest stems from the complementary functionalities of these approaches, and their respective strengths and weaknesses [99,100]. For example, neural networks excel at learning from vast amounts of raw, unstructured data, while symbolic AI techniques require structured data and do not scale well to large amounts of data. Nowadays, the combination of symbolic and neural AI tools and paradigms is identified by the term neurosymbolic AI [100]. The combination of ASP and neural networks makes no exception in this panorama [99].

This paper presents several integrations of ASP with machine learning and neural networks that we have explored and experimented with over the past few years. In particular, we report on:

1. An industrial application of ASP and neural network that addresses the compliance-checking of electrical control panels [23].
2. Two combinations of ASP with large language models for enabling reasoning from text [52,89].
3. Solver selection pipelines that leverage machine learning techniques to boost the evaluation of ASP programs [1,4,5].

Our experience reinforces the viability of these approaches, that have demonstrated their potential to effectively address complex challenges better than solutions only based on ASP. Furthermore, it suggests that such combinations will play an important role in a wide range of innovative AI applications in the future.

Paper Structure. The paper is structured as follows. Section 2 provides background notions on Answer Set Programming. Sections 3 describes the application of ASP in the problem of electric panel compliance; Sect. 4.3-4.2 address LLM-based approaches on translating natural language into ASP and question answering exploiting ASP background knowledge; Sect. 5 describes a machine learning-based algorithm selection approach for Quantified Answer Set Programming. We conclude the paper in Sect. 6 with discussion of related works and future perspectives for ASP in this domain.

2 Preliminaries

We first recall some basic notions of Answer Set Programming and Learning from Answer Sets, that will be mentioned in the remainder of the paper.

2.1 Answer Set Programming

In this section, we recall syntax and semantics of ASP.

ASP Syntax. In ASP a *term* is either a *constant* or a *variable*, where *variables* are strings starting with uppercase letters and *constants* are either integer numbers or strings starting with lowercase letters. *Atoms* are expressions of the form $p(t_1, \cdots, t_n)$ where p is a predicate of arity n and $t_1, \cdots, t_n$ are terms. An atom is *ground* if it contains no variable. A *literal* is either an atom a or its negation *not* a where *not* denotes negation as failure. Literals of the form a are said to be positive; whereas literals of the form *not* a are said to be negative. A *rule* is an expression of the form $h_1 \mid \ldots \mid h_m \leftarrow b_1, \cdots, b_n$ where $b_1, \cdots, b_n$ is a conjunction of literals, referred to as body, $n \geq 0$, $h_1, \ldots, h_m$ is a disjunction of atoms referred to as head, and $m \geq 0$. A *fact* is a rule with an empty body (i.e. $n = 0$). A *constraint* is a rule with an empty head (i.e. $m = 0$). *Weak constraints* are expression of the form $\leftarrow^w b_1, \ldots, b_n$ $[w@l, T]$, where $b_1, \ldots, b_n$ is a conjunction of literals, w, l, and T are terms. A rule, as well as a weak constraint, is said to be *safe* if every variable appears in at least one positive literal of the body. A *program* is a finite set of safe rules and weak constraints.

Answer Set Semantics. Given a program P, the *Herbrand Universe* U_P denotes the set of constants in P; the *Herbrand Base* B_P denotes the set of ground atoms that can be obtained from predicates in P and constants in U_P. Given a rule $r \in P$, we denote by $ground(r)$ the set of possible ground instantiations of r obtained by mapping variables in r with constants in U_P; and $ground(P)$ denotes the union of ground instantiations of rules in P. An *interpretation* $I \subseteq B_P$ is a set of atoms. Given an interpretation I, a positive (resp. negative) literal $l = a$ (resp. $l = not\ a$) is true w.r.t. I, if $a \in I$ (resp. $a \notin I$); otherwise l is false. A conjunction of literals L is true w.r.t. I, denoted by $I \models L$, if all the literals in L are true w.r.t. I; whereas a disjunction of literals L is true w.r.t. I, denoted by $I \models L$, if at least one literal of L is true w.r.t. I. An interpretation I is a *model* of P if for each rule $r \in ground(P)$, the head of r is true whenever the body of r is true. Given a program P and an interpretation I, the (Gelfond-Lifschitz) reduct [8], P^I, is the program obtained from $ground(P)$ by (i) removing all those rules having in the body a negative literal l that is false w.r.t. I, and (ii) removing negative literals from the body of remaining rules. Given a program P and a model I, I is a *stable model* or an *answer set* of P if there is no $I' \subset I$ such that I' is a model of P^I. The set of answer set of the program P is denoted by $AS(P)$. P is coherent if it admits at least one answer set, otherwise it is incoherent. Weak constraints do not affect the coherence of the programs but they define preferences among their answer sets. Thus, given $M \in AS(P)$, $ws(M, P)$ denote the set of weak constraint violations that is: $ws(M, P) = \{(w, l, T) \mid \leftarrow^w b_1, \ldots, b_n\ [w@l, T] \in ground(P)$ and $M \models b_1, \ldots, b_n\}$. Given an integer l, then $\mathcal{C}(M, P, l)$ denotes the cost of M at level l, that is $\mathcal{C}(M, P, l) = \sum_{(w,l,T) \in ws(M,P)} w$. Given two answer set of P, $M_1, M_2 \in AS(P)$, we say that M_1 is *dominated* by M_2 if there exits an integer l such that $\mathcal{C}(M_1, P, l) > \mathcal{C}(M_2, P, l)$ and for each integer $l' > l$, $\mathcal{C}(M_1, P, l') = \mathcal{C}(M_2, P, l')$. Let $M \in AS(P)$, then M is an optimal answer set if does not exist $M' \in AS(P)$ such that M is dominated by M'. The set of optimal answer set of the program P is denoted by $OptAS(P)$.

We refer the interested reader to the specific literature for a more detailed account [8,24].

2.2 Learning from Answer Sets

Learning from Answer Sets (LAS; [62]) is an inductive logic programming [73] framework based on ASP. Informally, a LAS learning task, consists of learning a logic program h, known as "inductive hypothesis" that together with a logic program B, known as "background knowledge", covers a set of examples E.

A *context-dependant example* e is a tuple $e = (e^{inc}, e^{exc}, e^{ctx})$, each component respectively known as *inclusion* set, *exclusion* set and *context* of the example e. e^{inc}, e^{exc} are sets of ground atoms, while e^{ctx} is a logic program. Given a background knowledge B and a hypothesis $h \subseteq H$, we say that the interpretation I *covers* the example e if there exists an answer set of $B \cup h \cup e_{ctx}$ such that $e^{inc} \subseteq I$, $e^{exc} \cup I = \{\}$[1].

A LAS *learning task* is a tuple (B, H, E) of background knowledge, hypothesis space and set of examples respectively.

Given a learning task (B, H, E) we say that the hypothesis $h \subseteq H$ is an *inductive solution* for the task if h covers each examples $e \in E$. Furthermore, the *cost* of h is the total number of literals in h. We say that h is *optimal* if no smaller $h' \subseteq H$ is an inductive solution for the task (B, H, E).

3 ASP and Neural Networks: Application to Compliance Checking

This section describes an industrial application of Answer Set Programming and Neurosymbolic AI for verifying electrical control panels compliance [23]. It was developed within the "MAP4ID—Multipurpose Analytics Platform 4 Industrial Data" project, which was partially founded by the Italian Ministry of Development.

3.1 Context: Electrical Control Panels Verification

In layman's terms, an *electrical control panel* (ECP) is a "box that contains electrical components", a container for devices such as sensors, switches, actuators, and circuits. ECPs are indeed ubiquitous in industrial complexes and factories.

ECP layouts are formally defined as schematics. Schematics provide the necessary information-in formal terms-to assemble the ECP and guarantee its effectiveness, safety, and high quality. They contain mainly information about components and connectivity between components. The problem of ECP compliance is verifying whether an ECP matches the formal specifications in a schematic.

Testing whether an ECP is compliant is far from trivial. It requires skilled technicians to inspect the ECP. Although this is often tedious, time-consuming, and error-prone, it is critical to establishing high-quality standards.

Hence, developing systems that automatically handle this kind of task is highly desirable for companies, as witnessed by several research projects and research lines addressing similar issues [14–17,19].

3.2 A Neurosymbolic Framework for ECPs Compliance

The objective of this project was to create a system to automatically detect anomalies and defects in ECPs (wrong type of components, misplaced components in the schematic, absent or redundant components, etc.) from pictures

---
[1] The LAS framework also allows for *negative* examples. The interested reader can refer to [62].

of assembled ECPs. Furthermore, the development had to cope with two challenges: the lack of design standards in ECPs and the lack of an adequate volume of high-quality, labeled training data.

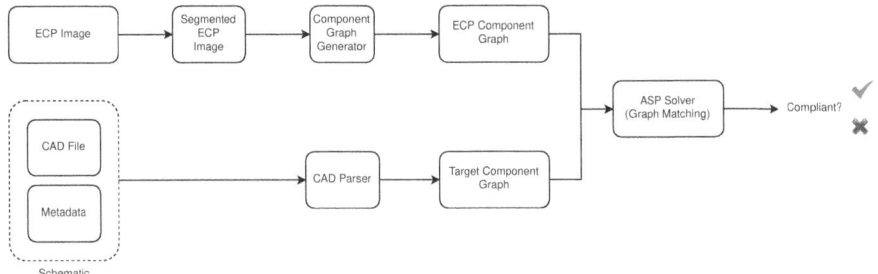

Fig. 1. Workflow for ECP compliance verification.

The proposed solution, depicted in Fig. 1, integrates a deep learning-based computer vision module with an ASP reasoning module. The computer vision module detects the location and type of each component that appears in the ECP, whereas the reasoning module deals with compliance-related reasoning aspects. The connection between the two modules consists of a parser that maps the object detection output into ASP facts.

First, input ECP images are scanned by the computer vision module (a pre-trained convolutional neural network [13,18]) to detect and segment objects of interest. The output of object detection and segmentation is then mapped into ASP facts, which define an *ECP component graph*. Similarly, the ECP schematic is parsed into ASP facts. The reasoning component, an ASP solver, checks the compliance of the ECP w.r.t. the given schematic, by solving a combinatorial problem that is, in its essence, similar to a graph matching problem.

The input to the system is the image of an ECP, its schematic, and other metadata. The computer vision module (a pre-trained convolutional neural network [13,18]) scans the ECP image to detect objects of interest. A translation module, namely "Component Graph Generator", translates the output of the computer vision module into a set of ASP facts, which define the ECP component graph. Likewise, a dedicated parser ("CAD Parser") translates the input schematic and metadata into ASP facts. The reasoning component, mainly a wrapper around an ASP solver, checks the compliance of the ECP with the given schematic by solving a combinatorial optimization problem that is intuitively a graph-matching problem augmented with domain knowledge about the ECP domain.

Figure 2 provides an excerpt of the logic program in the reasoning module, along with a brief comment on its role. We refer the reader to the original paper [23] for a thorough discussion of the encoding. The *object*/7 atoms encode the objects in the schematic and ECP and are generated by the CAD

Parser and Graph Component Generator modules. In particular the atom

$$object(type, id, x_0, y_0, x_1, y_1, m)$$

models a component with identifier id and type $type$ that belongs to the schematic ($m = $ "cad") or to the ECP ($m = $ "net") component graph, that is contained within the *bounding box* (that is, a rectangle that encloses the object) defined by the upper-left point $(x_0, y_0) \in \mathbb{N}^2$ and bottom-right point $(x_1, y_1) \in \mathbb{N}^2$.

```
1   % Guess a mapping between the CAD components and the ECP components.
2   previous(ID, Start_ID, D, M):- between(ID, Start_ID, _ , D, M).
3   after(ID, End_ID, D, M):- between(ID, _, End_ID, D, M).
4   simpObject(C1,ID1,M) :- object(C1,ID1,_,_,_,_,M).
5   mapped(ID1,ID2) || noMapped(ID1,ID2) :- simpObject(C1,ID1,"cad"),simpObject(C1,ID2,"net").
6
7   % No element in the graphs is mapped twice.
8   :- mapped(Cad_ID,Net_ID1), mapped(Cad_ID,Net_ID2), Net_ID1!=Net_ID2.
9   :- mapped(Cad_ID1,Net_ID), mapped(Cad_ID2,Net_ID), Cad_ID1!=Cad_ID2.
10
11  % Minimize the CAD elements without a mapping
12  atLeastOne(Cad_ID) :- mapped(Cad_ID,_).
13  :~ simpObject(C1,ID1,"cad"), not atLeastOne(ID1). [1@3,C1,ID1]
14
15  % Optimize mapping by relative position
16  :~ mapped(Cad_ID1, Net_ID1), mapped(Cad_ID2,Net_ID2), previous(Cad_ID1, Cad_ID2,DIR,"cad"), not previous(
        Net_ID1, Net_ID2, DIR,"net"). [1@2,Cad_ID1, Net_ID1,Cad_ID2,Net_ID2,DIR]
17  :~ mapped(Cad_ID1,Net_ID1), mapped(Cad_ID2,Net_ID2), after(Cad_ID1, Cad_ID2,DIR,"cad"), not after(Net_ID1,
        Net_ID2,DIR,"net"). [1@2,Cad_ID1,Net_ID1,Cad_ID2,Net_ID2,DIR]
18  :~ mapped(Cad_ID1, Net_ID1), previous(Cad_ID1, Cad_ID2, DIR,"cad"), absent(_,Cad_ID2). [1@2,Cad_ID1,Net_ID1,
        Cad_ID2,DIR]
19  :~ mapped(Cad_ID1, Net_ID1), after(Cad_ID1, Cad_ID2, DIR,"cad"), absent(_,Cad_ID2). [1@2,Cad_ID1,Net_ID1,
        Cad_ID2,DIR]
20
21  % Optimize mapping by distance
22  :~ mapped(Cad_ID, Net_ID), manhattan(Cad_ID, Net_ID, Dis,"cad","net"). [Dis@1,Cad_ID,Net_ID,Dis]
23
24  % Define missing and redundant components.
25  mappedCad(ID1):- mapped(ID1,_).
26  mappedNet(ID1):- mapped(_,ID1).
27  absent(C1,ID1):- simpObject(C1,ID1,"cad"), not mappedCad(ID1).
28  excess(C1,ID1):- simpObject(C1,ID1,"net"), not mappedNet(ID1).
```

Fig. 2. The logic program modeling compliance criteria for ECPs.

Decoupling the perception task from the reasoning task was pivotal for the project's success. Such a design choice enabled to address data shortage issues *at component detection level*, adopting data augmentation techniques to generate *synthetic ECPs*. On the other hand, addressing compliance reasoning with an ASP allows to update the system with new (or updated) compliance rules without further training of the object detection component, as well as cope with the lack of labeled data (that is, "which compliance issue affect a given ECP image") but rather exploit domain knowledge about ECPs and their schematics. A deep learning-only solution in this domain would have been less flexible, requiring re-training and an ad-hoc dataset to handle additions (or changes) to compliance rules—whereas in our solution it is sufficient to edit the logic program underlying the reasoning module, and no re-training of the computer vision component.

Overall, the proposed approach has shown acceptable performances, being both highly accurate (above 90%[2]) and fast (response times within seconds). Thus, the system met all the criteria for integration in a support tool to assist inspectors in their daily activities and fully satisfied the project requirements.

4 Combining ASP and LLMs: Two Ways

This section introduces some recent neurosymbolic approaches that integrate the strengths of Large Language Models with the expressive power of Answer Set Programming to tackle well-known challenges in the field of Natural Language Processing (NLP).

4.1 Large Language Models

The introduction of LLM models, such as GPT and BERT, has revolutionized Natural Language Processing (NLP) by enabling machines to process and generate human language with unprecedented accuracy. These deep neural network models owe their effectiveness to transformer-based architectures [32], which utilize self-attention mechanisms to process and contextualize vast amounts of text. Most currently available LLMs have billions of parameters and are trained in a self-supervised way to predict missing tokens or the next token in a given sequence. LLMs are usually instructed through text prompts to solve a specific task, such as translating languages or answering questions. They have also shown high performance in addressing semantic parsing tasks, i.e., converting text into a structured format for analysis [94, 96, 97].

4.2 ASP for LLM

A hot topic that elicits much interest in the Artificial Intelligence (AI) community is how to endow machines with commonsense and reasoning capabilities. Currently, most of the existing AI models used to address tasks such as machine comprehension and question and answering (Q&A), may do so by learning patterns and statistical features from the training data rather than learning commonsense knowledge or applying some general commonsense reasoning. This limitation causes these models to see their performance affected when facing tasks requiring some reasoning. The introduction of the LLMs greatly improved machines' capabilities to process, understand, and generate human language, thus boosting many Natural Language Processing (NLP) tasks. Despite these recent advances, several studies have found that LLMs struggle to provide

---

[2] The ASP component performs a crisp reasoning task with no uncertainty or errors in its output. Thus, the overall predictive error at inference time is determined solely by object detection and detection errors, due for example to overlapping components or images taken from an out-of-distribution perspective wrt the computer vision training dataset.

accurate answers [90] and underperform on natural language reasoning benchmarks [91,92]. In addition, the lack of explainability and transparency affecting most deep learning-based approaches makes it difficult to accurately verify whether these models learn or perform some commonsense reasoning [93].

Recent neurosymbolic approaches have addressed some of the limitations of LLMs by combining their generative capabilities with formal symbolic reasoning techniques. Nye et al. [94] proposed enhancing the coherence and consistency of LLM-generated story completions by employing an LLM-based semantic parser to map the text into formal representations, and then using a symbolic reasoner to check which of the LLM's generated completion sentences are correct. Ishay et al., in an effort to solve logic puzzles, proposed to combine LLMs with ASP [95]. On the other hand, Yang et al. combine a GPT3-based semantic parsing with a hand-crafted ASP knowledge module to handle Q&A tasks, showing good performance on several natural language reasoning benchmarks [96]. Although these approaches lead to robust reasoning from text, it is worth noting that much of the symbolic knowledge must be hand-crafted, meaning that addressing new problems will require some effort and time from humans.

In more recent work, Kareem et al. presented LLM2LAS [89], a system designed to address these limitations while preserving reasoning capacity. LLM2LAS learns common-sense knowledge for machine comprehension by combining an LLM-based semantic parser for generating symbolic representations of a story and questions, with ILASP[3], a tool for learning common-sense knowledge under the answer set semantics. The approach automatically generates an ILASP learning task from a given story, questions and answers, and iteratively learns relevant common-sense knowledge that can be used to answer questions about unseen texts. As illustrated in Fig. 3, the LLM2LAS architecture highlights two key components: (i) the LLM-based semantic parser, and (ii) the Learner.

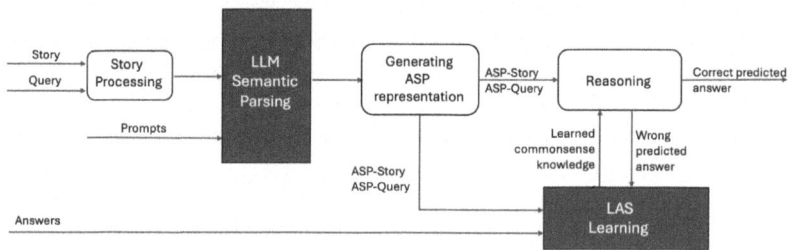

Fig. 3. Architecture of LLM2LAS [89].

The LLM-based semantic parser uses the *Few-shot prompting*[4] technique to ask an LLM model for the symbolic representations of the story sentences

---

[3] ILASP is a state-of-the-art implementation of the LAS framework.

[4] Few-shot prompting is a technique in which the model is given a few examples in the prompt to help it understand the task and improve output.

and question. Table 1 shows the semantic parsing results (second column) for a short story. On the other hand, when the Reasoner produces a wrong answer (or no answer), the system starts a common-sense learning process through the Learner. To this aim, the Learner creates a context-dependent learning task for ILASP that considers the correct and incorrect answers and the symbolic representations produced by the LLM-based parser. Although ILASP supports the explicit definition of the hypothesis space H, it is more common to *implicitly* define H by means of *mode biases*, that is, declarative expressions that define which kind of literals can belong to the head or body of each rule $\rho \in H$. A mode declaration is a literal, where the abstract arguments can be either var(t) or const(t), with t being a constant term referred to as a *type*. Informally, a literal is *compatible* with a mode declaration m if it can be constructed by replacing every instance of var(t) in m with a variable of type t, and every const(t) in m with a constant of type t. The set of constants of each type is given with the task, together with the maximum number of variables a rule can take.

Table 1. Sentences and symbolic representations for a short story [89].

Sentence	Symbolic Representation	Mode Bias Representation
Mice are afraid of wolves.	be_afraid_of(mouse,wolf).	be_afraid_of(nn,nn).
Mary is a mouse.	be(mary,mouse).	be(nnp,nn).
Cats are afraid of sheep.	be_afraid_of(cat,sheep).	be_afraid_of(nn,nn).
Winona is a mouse.	be(winona,mouse).	be(nnp,nn).
Wolves are afraid of cats.	be_afraid_of(wolf,cat).	be_afraid_of(nn,nn).
Emily is a mouse.	be(emily,mouse).	be(nnp,nn).
What is Mary afraid of?	be_afraid_of(mary,V1).	be_afraid_of(nnp, nn).

Specifically, LLM2LAS creates the mode declarations for the rules head by appealing to the question's symbolic representations, whereas for rules bodies, it relies on the non-question symbolic representations. The argument types for the mode bias constants and variables are determined using the POS tagging data of the sentences. Table 1 (third column) shows the mode bias representations for the example story. The inclusion and exclusion sets for the context-dependent examples are created using the correct and incorrect answers, respectively. The context consists of all representations of the sentences before the question that initiated the learning. Figure 4 illustrates the ILASP learning task created for the previous story. Note that the mode bias for the rule's head and body are specified using the (#modeh) and (#modeb) directives, respectively. For this example, ILASP learns the following rule:

be_afraid_of(V1,V2) :- be_afraid_of(V3,V2); be(V1,V3).

```
1   % Mode bias
2   #modeb(be_afraid_of(var(nn),var(nn))).
3   #modeh(be_afraid_of(var(nnp),var(nn))).
4   #modeb(be(var(nnp),var(nn))).
5   #maxv(3).
6   #max_penalty(50).
7
8   % Context-dependent examples
9   #pos({be_afraid_of(mary,wolf)},{},{
10    be(emily,mouse). be(mary,mouse).
11    be_afraid_of(wolf,cat). be(winona,mouse).
12    be_afraid_of(cat,sheep). be_afraid_of(mouse,wolf).
13  }).
```

Fig. 4. ILASP learning file for the short story [89].

which allows LLM2LAS to answer questions similar to *"What is Mary afraid of?"*, given that the information is available.

LLM2LAS showed good results on the bAbi [98] dataset, matching the state-of-the-art results but considerably reducing the human effort in coding the symbolic knowledge. The approach can be applied to larger and more complex datasets by leveraging LLM models with enhanced text understanding and improved semantic parsing capabilities.

4.3 LLM for ASP

Programming has always been a difficult task requiring a lot of time and expertise from people. Thanks to the recent advances in the NLP field, this situation is changing, largely due to the introduction of several tools that can assist users in the automatic composition of computer programs. However, most existing tools of this kind, like Copilot [46], primarily support widely used programming languages, with limited or no support for many problem-solving formalisms like Answer Set Programming (ASP). In the last few years, several works [40–43,51] have been focused on creating programming environments and tools to assist programmers in coding ASP, including advanced editors, debuggers, testing frameworks, and others. Still, coding in ASP remains highly challenging for beginners, particularly those with limited mathematical and logical backgrounds, and is also time-consuming for expert programmers. On the other hand, it is worth noting that generative AI tools such as ChatGPT [44,45,50] or Claude [48,49] can generate some ASP code snippets through appropriate prompts, but it is easy to demonstrate that these AI models struggle to produce valid and correct ASP programs. Thus, the challenge of reliably composing ASP programs from user requests in natural language continues to be a prominent and evolving research area.

In this context, several approaches have begun to take initial steps toward addressing this gap, as in the case of the NL2ASP tool [52]. The authors proposed a two-step architecture that uses the advances in the Neural Machine Translation (NMT) field to translate input sentences in natural language into a recently introduced controlled natural language (CNL) for ASP [86,87]. Then, the CNL statements are converted to ASP code using the CNL2ASP tool by Caruso et al. [87]. The overall system pipeline is depicted in Fig. 5.

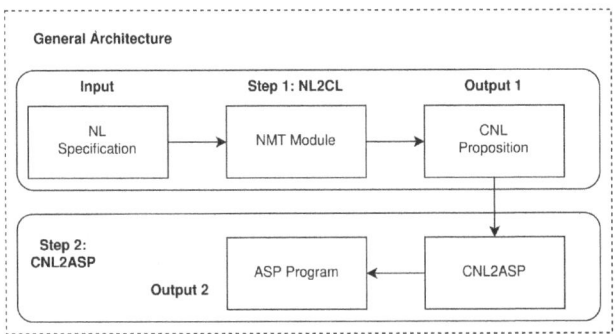

Fig. 5. Architecture for automatic composition of ASP programs.

In the first step, the approach experimented with BART [101] and T5 [47], two well-known encoder-decoder LLM models for machine translation[5]. Since these models were not specifically trained for the NL-to-CNL task, the authors introduced a new dataset called NL2CNL, intended to fine-tune (or train from scratch) these models. The dataset focused on graph-related problem specifications and was created following the workflow depicted in Fig. 6. The dataset creation followed a template-based semi-automatic method that started by collecting ASP programs from several sources. Then, each rule in a program was manually associated with the corresponding CNL and NL statements. These CNL-NL pairs facilitated the creation of statement templates by substituting specific parts of speech (e.g., nouns, verbs) with placeholders. The templates were expanded by replacing the placeholders with words selected from a pre-constructed *bag-of-words*. The resulting dataset was then further expanded by rephrasing each sentence five times using the *"text-davinci-003"* model by OpenAI. The aim of this last step was not just to increase the dataset size but also to improve its quality and diversity by introducing varied linguistic representations without losing the original meaning.

The system was evaluated based on the quality of the statements generated at each step. The NMT module demonstrated robust behavior in NL to CNL translation and exhibited a strong ability to generate syntactically correct CNL statements. On the other hand, the system was tested end-to-end, demonstrating good performance (19 out of 21 correct solutions) in generating ASP code for unseen problems, with only minor issues that required minimal

---

[5] T5 and BART belongs to the encoder-decoder transformer LLMs family.

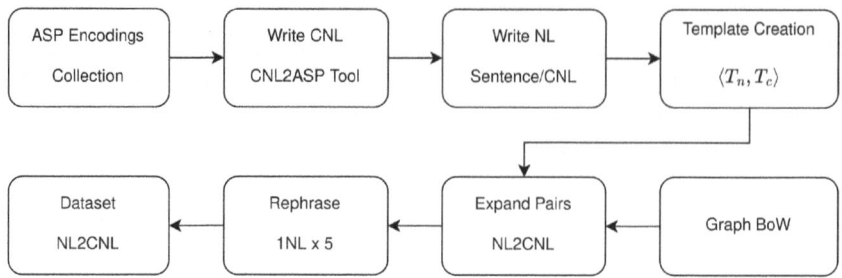

Fig. 6. NL2CNL Dataset creation workflow [52].

effort to fix. Another important experiment conducted in this work showed how general-purpose LLMs, such as ChatGPT, underperform in generating good ASP programs, even when prompted several times. Specifically, ChatGPT was asked for the ASP encodings for the same 21 problems used in the end-to-end experiment, mistaking most of the time (17/21). Nevertheless, whether these models can be used reliably for this task remains an open topic.

5 Neurosymbolic AI for Solving: Boosting Evaluation of Logic Programs

The efficient evaluation of logic programs plays a pivotal role in the success and widespread adoption of declarative formalism [10]. Over the years, significant efforts have been devoted to designing efficient evaluation techniques for ASP programs, but it is rare for a single technique to be universally effective for all types of ASP programs.

To this end, the task of predicting the most suitable solving technique has been extensively studied within the logic programming community, and machine learning methods have proved to be particularly effective [2–4]. Specifically, this task can be framed as a *multinomial classification problem*, where the target label represents possible symbolic reasoners, and the input features describe the syntactic properties of the program to be solved. Thus, equipping symbolic reasoners with neural components that learn which solving methodology is the most suitable for the domain of interest can provide a significant boost in performance.

Recently, such kind of methodologies have been proposed in the realm of Answer Set Programming with Quantifiers (ASP(Q)) [6]. The ASP(Q) system PyQASP [5] has been equipped with an algorithm selection methodology inspired by the ME-ASP approach [4], that automatically selects the most suitable back-end solver. As shown in Fig. 7, PyQASP^{Auto} demonstrated remarkable accuracy, achieving performance nearly identical to the virtual best solver. Furthermore, it solved significantly more instances than the available fixed back-end version of PyQASP, thereby significantly advancing the state-of-the-art in ASP(Q) solving.

In the realm of Answer Set Programming, instead, novel program evaluation techniques are emerging. One notable development is compilation-based ASP

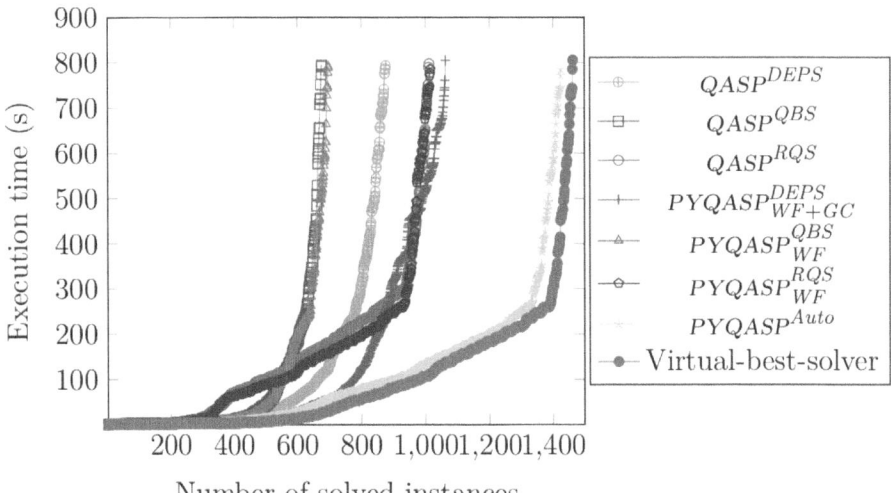

Fig. 7. Improvements in ASP(Q) evaluation

solving, proposed to address the grounding bottleneck [12,20] that affects current state-of-the-art ASP systems.

This approach involves compiling a logic program into an ad-hoc solver composed of rule-specific propagators that simulate the presence of ground rules during solving-without explicitly materializing them [7,9,11].

Recent advancements in this area have demonstrated substantial performance improvements by combining the standard Ground&Solve approach, employed by state-of-the-art systems, with compilation-based techniques [1]. Specifically, each rule of an ASP program can be either compiled into an ad-hoc grounding procedure or transformed into ad-hoc propagators, thus enabling an exponential number of solving strategies for a specific input program.

Experimental evaluations have shown that different blending strategies excel in different domains, and so deciding which rules should be grounded and which should be transformed into propagators significantly impacts the solving process.

This development opens the door to the design of novel algorithm selection methodologies tailored for compilation-based ASP solving. However, existing techniques cannot be directly applied here, as they rely on ground programs, whereas the objective of compilation-based methods is to avoid the grounding phase. This introduces not-trivial challenges but offers highly promising rewards.

6 Related Works

Answer Set Programming, with its knowledge representation and reasoning roots, has proven effective in various applications [88] involving different kinds of knowledge and reasoning tasks. Thus, Answer Set Programming seems a natural, yet currently underrepresented, target language for enabling neurosymbolic integration. We report in the following an analyiss of some related approaches.

Reasoning over Perception. The Neural|Symbolic pattern, architectures where symbolic solvers take as input appropriately-encoded neural networks' outputs, is the most intuitive approach to neurosymbolic integration. To date, most neurosymbolic applications fall in this category. In this setting, ASP has been applied to visual question answering [59–61] and the aforementioned compliance checking use case in industrial settings [23]. *Semantic parsing*-based approaches [102] that exploit LLMs to translate natural language into a set of facts to reason over with ASP can also fit this category [26,89].

Translating Natural Language into Logic Programs. Despite being a popular formalism for knowledge representation and combinatorial optimization, in *absolute terms* only a few of developers possess working knowledge of Answer Set Programming. Thus, automatic translation of natural language specifications into ASP is an interesting problem to *bring ASP to the masses*. However, the unavailability of large amounts of data makes classic neural-machine translation techniques [57] ineffective in translating natural language into ASP specifications. In fact, to date, most of the publicly available ASP code consists of students' exercises or coding competition-like solutions—which often are not accompained by natural language explanations of the underlying principles in the encoding. Indeed, there are some works on translating natural language specifications directly into ASP programs [89,95], or intermediate representations, such as controlled natural language [86,87].

ML-Augmented ASP Systems. A popular instantiation of the Symbolic[Neural] pattern, primarily investigated in the realm of CSP, SAT, and automated planning, is learning solving heuristics from instance data or past solvers' executions. To date, the issue has not been explored for ASP, with due exception of [85] in the graph coloring domain and [84] for *negative two-literal programs* [82,83]. As current state-of-the-art ASP systems revolve around the ground-and-solve approach, similar considerations can be made for rewritings and heuristics during the grounding phase. Reference [81] investigates ML-driven grounding heuristics exploiting syntactic features of non-ground logic programs. We are unaware of works pursuing this aspect in ASP. However it has been attempted in SAT [75], CSPs [28,79], other combinatorial optimization formalisms [29,76–78] and planning [72,74]. This research stream also encompasses the yet unexplored question of *how should a logic program be represented* to be effectively processed by a neural network, especially if one wishes to do so at the non-ground level. In this area, *suitable* graph representations—preserving semantic properties— for logic programs were investigated in [27,80].

Semantic Losses and Neural-Predicates Extensions of ASP. The seminal work [71] introduces the notion of *semantic loss*, which refers to translating logical constraints, expressed as a propositional formula, into a differentiable loss term for neural networks. From the technical point of view, the underlying machinery that enables this is *knowledge compilation* [69,70] and *weighted model counting* [55,58]. The advantages of ASP in KR&R wrt SAT are well established in the literature. Recent advancements in model counting and knowledge compilation for ASP [68] can provide the opportunity to lift SAT-based techniques to ASP, providing a richer language to express semantic losses. Similarly, linear algebraic techniques [30,31,67] look promising in this regard, providing an alternative— "differentiable by design"— way to inject logical knowledge at training time as well as scaling up end-to-end ASP computation [35]. Neural-predicates extensions of ASP [25,38,39], that is, approaches that extend the base language with *neural predicates* in DeepProbLog [37] fashion, have also been explored, and can play a similar role to semantic loss in being able to *regularize* neural networks' learning with respect to a background knowledge expressed as a logic program.

Detecting and Repairing Inconsistencies in LLMs Reasoning. Hallucinations are most likely the main issues hindering real-world LLMs applications [34]. Although different prompting techniques, such as *chain of thought* [91] and *chain of verification* [56], can mitigate the issue, such techniques do not currently *solve* the issue, which appears to be structural in the Transformer [32] architecture [33]. ASP has a long-standing tradition of dealing with partial knowledge and inconsistency as witnessed by applications in paracoherent reasoning [63] and consistent query answering [64]. Due to technical advancements in *minimal unsatisfiable cores* enumeration [65,66], paired with ASP flexibility in representing different types of knowledge (as an example, dealing with both spatial reasoning [36], temporal reasoning [53,54]) makes ASP an appealing tool to both *detect inconsistencies* and possibly *reconcile* and *explain* LLM outputs involving well-defined, ASP-expressible knowledge domains.

7 Conclusion

ASP is a well-known symbolic AI formalism developed in the area of knowledge representation and reasoning. The declarative nature of ASP makes it an ideal candidate for modeling and solving AI reasoning tasks. The combination of ASP with modern AI systems based on subsymbolic techniques has a huge potential deriving from the possibility of combining the strengths of ASP with the ones of neural networks, which will give rise to several possible neurosymbolic approaches to AI. In this paper, we have described some of them, including (*i*) an industrial application of ASP and computer vision to address the compliance of electrical control panels, (*ii*) two possible combinations of ASP with large language models for enabling reasoning from text, and (*iii*) the usage of machine learning techniques to boost the evaluation of ASP programs by implementing solver selection pipelines.

These experiences confirm the viability of these combinations, and outline some weaknesses that can be the venue for future investigations. For example, although recent advances have made strides in automatic program generation, reliably producing correct ASP code from natural language remains challenging. Also, using neural networks to boost solver performance that, in our experience, was limited to algorithm selection, should be subject to additional efforts. Concerning applications, simple neurosymbolic pipelines can also be effective in practice, but finding more integrated solutions can further push the boundaries of the applicability of ASP.

Acknowledgement. This work was partially supported by the Italian Ministry of Research (MUR) under PRIN project PINPOINT - CUP H23C22000280006; and by the PNRR projects FAIR "Future AI Research" - Spoke 9 - WP9.1 - CUP H23C22000860006, and The Italian Ministry of Enterprise Industry and Made in Italy (MIMIT) under project EI-TWIN n. F/310168/05/X56 CUP B29J24000680005.

References

1. Dodaro, C., Mazzotta, G., Ricca, F.: Blending grounding and compilation for efficient ASP solving. In: Proceedings of the 21st International Conference on Principles of Knowledge Representation and Reasoning, pp. 317–328 (2024). https://doi.org/10.24963/kr.2024/30
2. Gebser, M., Kaminski, R., Kaufmann, B., Schaub, T., Schneider, M.T., Ziller, S.: A portfolio solver for answer set programming: preliminary report. In: Delgrande, J.P., Faber, W. (eds.) LPNMR 2011. LNCS (LNAI), vol. 6645, pp. 352–357. Springer, Heidelberg (2011). https://doi.org/10.1007/978-3-642-20895-9_40
3. Balduccini, M.: Learning and using domain-specific heuristics in ASP. AI Commun. **24**, 147–164 (2011)
4. Maratea, M., Pulina, L., Ricca, F.: A multi-engine approach to answer-set programming. Theory Pract. Log. Program. **14**, 841–868 (2014)
5. Faber, W., Mazzotta, G., Ricca, F.: An efficient solver for ASP(Q). Theory Pract. Log. Program. **23**, 948–964 (2023)
6. Amendola, G., Ricca, F., Truszczynski, M.: Beyond NP: quantifying over answer sets. TPLP **19**, 705–721 (2019). https://doi.org/10.1017/S1471068419000140
7. Mazzotta, G., Ricca, F., Dodaro, C.: Compilation of aggregates in ASP systems. In: AAAI, pp. 5834–5841. AAAI Press (2022)
8. Gelfond, M., Lifschitz, V.: Classical negation in logic programs and disjunctive databases. New Gener. Comput. **9**, 365–386 (1991)
9. Dodaro, C., Mazzotta, G., Ricca, F.: Compilation of tight ASP programs. In: ECAI. Frontiers in Artificial Intelligence and Applications, vol. 372, pp. 557–564. IOS Press (2023)
10. Gebser, M., Leone, N., Maratea, M., Perri, S., Ricca, F., Schaub, T.: Evaluation techniques and systems for answer set programming: a survey. In: IJCAI, pp. 5450–5456. ijcai.org. (2018)
11. Cuteri, B., Dodaro, C., Ricca, F., Schüller, P.: Overcoming the Grounding bottleneck due to constraints in ASP solving: constraints become propagators. In: IJCAI, pp. 1688–1694. ijcai.org. (2020)
12. Ostrowski, M., Schaub, T.: ASP modulo CSP: the clingcon system. Theory Pract. Log. Program. **12**, 485–503 (2012)

13. Ren, S., He, K., Girshick, R., Sun, J.: Faster R-CNN: towards real-time object detection with region proposal networks. In: Proceedings of NIPS - Volume 1, pp. 91-99. MIT Press (2015)
14. Subakti, H., Jiang, J.-R.: Indoor augmented reality using deep learning for industry 4.0 smart factories. In: 2018 IEEE COMPSAC, vol. 02, pp. 63–68. (2018). https://doi.org/10.1109/COMPSAC.2018.10204
15. Villalba-Diez, J., Schmidt, D., Gevers, R., Meré, J., Buchwitz, M., Wellbrock, W.: Deep learning for industrial computer vision quality control in the printing industry 4.0. Sensors **19**, 3987 (2019). https://doi.org/10.3390/s19183987
16. Banus Paradell, N., Boada, I., Xiberta, P., Toldra, P., Bustins, N.: Deep learning for the quality control of thermoforming food packages. Sci. Reports **11**, 21887 (2021). https://doi.org/10.1038/s41598-021-01254-x
17. Schmitt, J., Bönig, J., Borggräfe, T., Beitinger, G., Deuse, J.: Predictive model-based quality inspection using machine learning and edge cloud computing. Adv. Eng. Inform. **45**, 101101 (2020). https://doi.org/10.1016/j.aei.2020.101101
18. LeCun, Y., Bengio, Y., Hinton, G.: Deep learning. Nature 521, 436–444 (2015)
19. Tanuska, P., Spendla, L., Kebisek, M., Duris, R., Stremy, M.: Smart anomaly detection and prediction for assembly process maintenance in compliance with industry 4.0. Sensors **21**, 2376 (2021)
20. Calimeri, F., Gebser, M., Maratea, M., Ricca, F.: Design and results of the fifth answer set programming competition. Artif. Intell. **231**, 151–181 (2016). https://doi.org/10.1016/j.artint.2015.09.008
21. Erdem, E., Öztok, U.: Generating explanations for biomedical queries. Theory Pract. Log. Program. **15**, 35–78 (2015). https://doi.org/10.1017/S1471068413000598
22. Grasso, G., Iiritano, S., Leone, N., Ricca, F.: Some DLV applications for knowledge management. In: Erdem, E., Lin, F., Schaub, T. (eds.) LPNMR 2009. LNCS (LNAI), vol. 5753, pp. 591–597. Springer, Heidelberg (2009). https://doi.org/10.1007/978-3-642-04238-6_63
23. Barbara, V., et al.: Neuro-symbolic AI for compliance checking of electrical control panels. Theory Pract. Log. Program. **23**, 748–764 (2023)
24. Brewka, G., Eiter, T., Truszczynski, M.: Answer set programming at a glance. Commun. ACM **54**, 92–103 (2011). https://doi.org/10.1145/2043174.2043195
25. Yang, Z., Ishay, A., Lee, J.: NeurASP: embracing neural networks into answer set programming. In: Proceedings of the Twenty-Ninth International Joint Conference on Artificial Intelligence, IJCAI, pp. 1755–1762. ijcai.org. (2020). https://doi.org/10.24963/ijcai.2020/243
26. Alviano, M., Grillo, L.: Answer set programming and large language models interaction with YAML: preliminary report. In: CILC. CEUR Workshop Proceedings, vol. 3733. CEUR-WS.org. (2024)
27. Konczak, K., Linke, T., Schaub, T.: Graphs and colorings for answer set programming. Theory Pract. Log. Program. **6**, 61–106 (2006)
28. Tönshoff, J., Kisin, B., Lindner, J., Grohe, M.: One model, any CSP: graph neural networks as fast global search heuristics for constraint satisfaction. In: IJCAI, pp. 4280–4288. ijcai.org. (2023)
29. Gasse, M., et al.:. the machine learning for combinatorial optimization competition (ML4CO): results and insights. CoRR. Vol. abs/2203.02433 (2022)
30. Nguyen, H.D., Sakama, C., Sato, T., Inoue, K.: Computing logic programming semantics in linear algebra. In: Kaenampornpan, M., Malaka, R., Nguyen, D.D., Schwind, N. (eds.) MIWAI 2018. LNCS (LNAI), vol. 11248, pp. 32–48. Springer, Cham (2018). https://doi.org/10.1007/978-3-030-03014-8_3

31. Aspis, Y., Broda, K., Russo, A., Lobo, J.: Stable and supported semantics in continuous vector spaces. In: KR, pp. 59–68 (2020)
32. Vaswani, A., et al.: Attention is all you need. In: NIPS, pp. 5998–6008 (2017)
33. Peng, B., Narayanan, S., Papadimitriou, C.H.: On Limitations of the Transformer Architecture. CoRR. abs/2402.08164 (2024)
34. Sahoo, P., Meharia, P., Ghosh, A., Saha, S., Jain, V., Chadha, A.: A comprehensive survey of hallucination in large language, image, video and audio foundation models. In: EMNLP (Findings), pp. 11709–11724. Association for Computational Linguistics (2024)
35. Sato, T., Takemura, A., Inoue, K.: Towards end-to-end ASP computation. CoRR. abs/2306.06821 (2023)
36. Baryannis, G., et al.: A trajectory calculus for qualitative spatial reasoning using answer set programming. Theory Pract. Log. Program. **18**, 355–371 (2018)
37. Manhaeve, R., Dumancic, S., Kimmig, A., Demeester, T., De Raedt, L.: Neural probabilistic logic programming in DeepProbLog. Artif. Intell. **298**, 103504 (2021)
38. Geh, R.L., Gonçalves, J., Silveira, I.C., Mauá, D.D., Cozman, F.G.: PASP: A Comprehensive Differentiable Probabilistic Answer Set Programming Environment For Neurosymbolic Learning and Reasoning. CoRR. abs/2308.02944 (2023)
39. Skryagin, A., Ochs, D., Dhami, D.S., Kersting, K.: Scalable neural-probabilistic answer set programming. J. Artif. Intell. Res. **78**, 579–617 (2023)
40. Alviano, M., Cirimele, D., Reiners, L.A.R.: Introducing ASP recipes and ASP chef. In: ICLP Workshops. CEUR, vol. 3437. CEUR-WS.org (2023)
41. Hahn, S., Sabuncu, O., Schaub, T., Stolzmann, T.: Clingraph: A System for ASP-based Visualization. CoRR. abs/2303.10118 (2023). https://doi.org/10.48550/ARXIV.2303.10118
42. Febbraro, O., Reale, K., Ricca, F.: ASPIDE: integrated development environment for answer set programming. In: Delgrande, J.P., Faber, W. (eds.) LPNMR 2011. LNCS (LNAI), vol. 6645, pp. 317–330. Springer, Heidelberg (2011). https://doi.org/10.1007/978-3-642-20895-9_37
43. Busoniu, P.-A., Oetsch, J., Pührer, J., Skocovsky, P., Tompits, H.: SeaLion: an eclipse-based IDE for answer-set programming with advanced debugging support. Theory Pract. Log. Program. **13**, 657–673 (2013). https://doi.org/10.1017/S1471068413000410
44. OpenAI. GPT-4 Technical report. arXiv preprint: arXiv:2303.08774 (2023)
45. Long, O., et al.: Training language models to follow instructions with human feedback. arXiv preprint arXiv:2203.02155 (2022)
46. Chen, M., et al.: Evaluating Large Language Models Trained on Code. CoRR. abs/2107.03374 (2021)
47. Raffel, C., et al.: Exploring the limits of transfer learning with a unified text-to-text transformer. J. Mach. Learn. Res. **21**, 1–67 (2020). http://jmlr.org/papers/v21/20-074.html
48. Anthropic: Model Card and Evaluations for Claude Models (2023). https://www.anthropic.com/news/claude-2
49. Anthropic: Introducing the next generation of Claude (2024). https://www.anthropic.com/news/claude-3-family
50. Brown, T., et al.: Language models are few-shot learners. In: Advances in Neural Information Processing Systems, vol. 33, pp. 1877–1901. Curran Associates, Inc. (2020). https://proceedings.neurips.cc/paper_files/paper/2020/file/1457c0d6bfcb4967418bfb8ac142f64a-Paper.pdf
51. De Vos, M., Kisa, D.G., Oetsch, J., Pührer, J., Tompits, H.: Annotating answer-set programs in Lana. Theory Pract. Log. Program. **12**, 619–637 (2012)

52. Santana, M.A.B., Kareem, I., Ricca, F.: Towards automatic composition of ASP programs from natural language specifications. In: IJCAI, pp. 6198–6206. ijcai.org. (2024)
53. Fionda, V., Ielo, A., Ricca, F.: LTLf2ASP: LTLf bounded satisfiability in ASP. In: Dodaro, C., Gupta, G., Martinez, M.V. (eds.) Logic Programming and Nonmonotonic Reasoning. LPNMR 2024. LNCS, vol. 15245, pp. 373–386. Springer, Cham (2024). https://doi.org/10.1007/978-3-031-74209-5_28
54. Brenton, C., Faber, W., Batsakis, S.: Answer set programming for qualitative spatio-temporal reasoning: methods and experiments. ICLP Techn. Commun. OASIcs. **52**, 4:1-4:15 (2016)
55. Gomes, C.P., Sabharwal, A., Selman, B.: Model counting. In: Handbook of Satisfiability. Frontiers in Artificial Intelligence and Applications, vol. 336, pp. 993–1014. IOS Press (2021)
56. Dhuliawala, S., et al.: Chain-of-verification reduces hallucination in large language models. In: ACL (Findings), pp. 3563–3578. Association for Computational Linguistics (2024)
57. Ranathunga, S., Lee, E.-S.A., Skenduli, M.P., Shekhar, R., Alam, M., Kaur, R.: Neural machine translation for low-resource languages: a survey. ACM Comput. Surv. **55**, 229:1-229:37 (2023)
58. Chakraborty, S., Meel, K.S., Vardi, M.Y.: Approximate model counting. In: Handbook of Satisfiability. Frontiers in Artificial Intelligence and Applications, vol. 336, pp. 1015–1045. IOS Press (2021)
59. Eiter, T., Ruiz, N.H., Oetsch, J.: A Modular neurosymbolic approach for visual graph question answering. In: NeSy. CEUR Workshop Proceedings, vol. 3432, pp. 139–149. CEUR-WS.org (2023)
60. Eiter, T., Geibinger, T., Higuera, N., Oetsch, J.: A logic-based approach to contrastive explainability for neurosymbolic visual question answering. In: IJCAI, pp. 3668–3676. ijcai.org. (2023)
61. Eiter, T., Higuera, N., Oetsch, J., Pritz, M.: A neuro-symbolic ASP pipeline for visual question answering. Theory Pract. Log. Program. **22**, 739–754 (2022)
62. Law, M.: Conflict-driven inductive logic programming. Theory Pract. Log. Program. **23**, 387–414 (2023)
63. Amendola, G., Dodaro, C., Faber, W., Ricca, F.: Paracoherent answer set computation. Artif. Intell. **299**, 103519 (2021)
64. Manna, M., Ricca, F., Terracina, G.: Consistent query answering via asp from different perspectives: theory and practice. Theory Pract. Log. Program. **13**, 227–252 (2013)
65. Alviano, M., Dodaro, C., Fiorentino, S., Previti, A., Ricca, F.: Enumeration of minimal models and MUSes in WASP. In: Gottlob, G., Inclezan, D., Maratea, M. (eds.) Logic Programming and Nonmonotonic Reasoning. LPNMR 2022. LNCS, vol. 13416, pp. 29–42. Springer, Cham (2022). https://doi.org/10.1007/978-3-031-15707-3_3
66. Alviano, M., Dodaro, C., Fiorentino, S., Previti, A., Ricca, F.: ASP and subset minimality: enumeration, cautious reasoning and MUSes. Artif. Intell. **320**, 103931 (2023)
67. Nguyen, T., Inoue, K., Sakama, C.: Linear algebraic partial evaluation of logic programs. In: NMR. CEUR Workshop Proceedings, vol. 3835, pp. 14–23. CEUR-WS.org (2024)
68. Eiter, T., Hecher, M., Kiesel, R.: aspmc: new frontiers of algebraic answer set counting. Artif. Intell. **330**, 104109 (2024)

69. Darwiche, A., Marquis, P.: A knowledge compilation map. J. Artif. Intell. Res. **17**, 229–264 (2002)
70. Darwiche, A., Marquis, P.: Knowledge compilation. Ann. Math. Artif. Intell. **92**, 1007–1011 (2024)
71. Xu, J., Zhang, Z., Friedman, T., Liang, Y., Van den Broeck, G.: A semantic loss function for deep learning with symbolic knowledge. In: ICML. Proceedings of Machine Learning Research, vol. 80, pp. 5498–5507. PMLR (2018)
72. Chen, D.Z., Thiébaux, S., Trevizan, F.W.: Learning domain-independent heuristics for grounded and lifted planning. In: AAAI, pp. 20078–20086. AAAI Press (2024)
73. Cropper, A., Dumancic, S.: Inductive logic programming at 30: a new introduction. J. Artif. Intell. Res. **74**, 765–850 (2022)
74. Micheli, A., Valentini, A.: Synthesis of search heuristics for temporal planning via reinforcement learning. In: AAAI, pp. 11895–11902. AAAI Press (2021)
75. Fournier, T., et al.: A deep reinforcement learning heuristic for SAT based on antagonist graph neural networks. In: ICTAI, pp. 1218–1222. IEEE (2022)
76. Marty, T., François, T., Tessier, P., Gautier, L., Rousseau, L.M., Cappart, Q.: Learning a generic value-selection heuristic inside a constraint programming solver. In: CP. LIPIcs, vol. 280, pp. 25:1–25:19. Schloss Dagstuhl - Leibniz-Zentrum für Informatik (2023)
77. Cappart, Q., Chételat, D., Khalil, E.B., Lodi, A., Morris, C., Velickovic, P.: Combinatorial optimization and reasoning with graph neural networks. J. Mach. Learn. Res. **24**, 130:1-130:61 (2023)
78. Boisvert, L., Verhaeghe, H., Cappart, Q.: Towards a generic representation of combinatorial problems for learning-based approaches. In: Dilkina, B. (ed.) Integration of Constraint Programming, Artificial Intelligence, and Operations Research. CPAIOR 2024. LNCS, vol. 14742, pp. 99–108. Springer, Cham (2024). https://doi.org/10.1007/978-3-031-60597-0_7
79. Pereira, G.: Civic poverty and recognition. In: Schweiger, G. (ed.) Poverty, Inequality and the Critical Theory of Recognition. PP, vol. 3, pp. 83–106. Springer, Cham (2020). https://doi.org/10.1007/978-3-030-45795-2_4
80. Linke, T., Sarsakov, V.: Suitable graphs for answer set programming. In: Baader, F., Voronkov, A. (eds.) LPAR 2005. LNCS (LNAI), vol. 3452, pp. 154–168. Springer, Heidelberg (2005). https://doi.org/10.1007/978-3-540-32275-7_11
81. Mastria, E., Zangari, J., Perri, S., Calimeri, F.: A machine learning guided rewriting approach for ASP logic programs. ICLP Techn. Commun. EPTCS. **325**, 261–267 (2020)
82. Wang, K., Wen, L., Kedian, M.: Random logic programs: linear model. Theory Pract. Log. Program. **15**, 818–853 (2015)
83. Namasivayam, G., Truszczyński, M.: Simple random logic programs. In: Erdem, E., Lin, F., Schaub, T. (eds.) LPNMR 2009. LNCS (LNAI), vol. 5753, pp. 223–235. Springer, Heidelberg (2009). https://doi.org/10.1007/978-3-642-04238-6_20
84. Ielo, A., Ricca, F.: Answer set computation of negative two-literal programs based on graph neural networks: preliminary results. In: CILC CEUR Workshop Proceedings, vol. 3002, pp. 188–195. CEUR-WS.org. (2021)
85. Dodaro, C., Ilardi, D., Oneto, L., Ricca, F.: Deep learning for the generation of heuristics in answer set programming: a case study of graph coloring. In: Gottlob, G., Inclezan, D., Maratea, M. (eds) Logic Programming and Nonmonotonic Reasoning. LPNMR 2022. LNCS, vol. 13416, pp. 145–158. Springer, Cham (2022). https://doi.org/10.1007/978-3-031-15707-3_12

86. Bruno, P., Caruso, S., Dodaro, C., Maratea, M.: A tool for reasoning over CNL sentences with temporal constructs. In: Datalog. CEUR Workshop Proceedings, vol. 3801, pp. 30–42. CEUR-WS.org. (2024)
87. Caruso, S., Dodaro, C., Maratea, M., Mochi, M., Riccio, F.: CNL2ASP: converting controlled natural language sentences into ASP. Theory Pract. Log. Program. **24**, 196–226 (2024)
88. Falkner, A.A., Friedrich, G., Schekotihin, K., Taupe, R., Teppan, E.C.: Industrial applications of answer set programming. Künstliche Intell. **32**, 165–176 (2018)
89. Kareem, I., Gallagher, K., Borroto, M., Ricca, F., Russo, A.: Using learning from answer sets for robust question answering with LLM. In: Dodaro, C., Gupta, G., Martinez, M.V. (eds.) Logic Programming and Nonmonotonic Reasoning. LPNMR 2024. LNCS, vol. 15245, pp. 112–125. Springer, Cham (2024). https://doi.org/10.1007/978-3-031-74209-5_9
90. Zheng, S., Huang, J., Chang, K.C.-C.: Why Does ChatGPT Fall Short in Answering Questions Faithfully? CoRR. abs/2304.10513. (2023)
91. Wei, J., et al.: Chain-of-thought prompting elicits reasoning in large language models. In: NeurIPS (2022)
92. Lake, B.M., Murphy, G.L.: Word meaning in minds and machines. CoRR. abs/2008.01766 (2020)
93. Sap, M., Shwartz, V., Bosselut, A., Choi, Y., Roth, D.: Commonsense reasoning for natural language processing. In: ACL (Tutorial), pp. 27–33. ACL. (2020)
94. Nye, M.I., Tessler, M.H., Tenenbaum, J.B., Lake, B.M.: Improving Coherence and consistency in neural sequence models with dual-system, neuro-symbolic reasoning. In: NeurIPS, pp. 25192–25204 (2021)
95. Ishay, A., Yang, Z., Lee, J.: Leveraging large language models to generate answer set programs. In: KR, pp. 374–383 (2023)
96. Yang, Z., Ishay, A., Lee, J.: Coupling large language models with logic programming for robust and general reasoning from text. In: ACL (Findings), pp. 5186–5219. Association for Computational Linguistics (2023)
97. Drozdov, A., et al.: Compositional semantic parsing with large language models. In: ICLR. OpenReview.net. (2023)
98. Weston, J., Bordes, A., Chopra, S., Mikolov, T.: Towards AI-complete question answering: a set of prerequisite toy tasks. In: ICLR (Poster) (2016)
99. Marra, G., Dumancic, S., Manhaeve, R., De Raedt, L.: From statistical relational to neurosymbolic artificial intelligence: a survey. Artif. Intell. **328**, 104062 (2024)
100. ten Teije, A., van Harmelen, F.: Architectural patterns for neuro-symbolic AI. In: Compendium of Neurosymbolic Artificial Intelligence. Frontiers in Artificial Intelligence and Applications, vol. 369, pp. 64–76. IOS Press (2023)
101. Lewis, M., et al.: BART: Denoising sequence-to-sequence pre-training for natural language generation, translation, and comprehension. In: Jurafsky, D., Chai, J., Schluter, N., Tetreault, J. (eds.) Proceedings of the 58th Annual Meeting of the Association for Computational Linguistics, pp. 7871–7880. Association for Computational Linguistics, Online, July 2020. https://doi.org/10.18653/v1/2020.acl-main.703
102. Gardner, M., Dasigi, P., Iyer, S., Suhr, A., Zettlemoyer, L.: Neural semantic parsing. In: Artzi, Y., Eisenstein, J. (eds.) Proceedings of the 56th Annual Meeting of the Association for Computational Linguistics: Tutorial Abstracts, Association for Computational Linguistics, Melbourne, Australia, July 2018, pp. 17–18 (2018). https://doi.org/10.18653/v1/P18-5006

Understanding Artificial Intelligence in Chess: The RubiChess Case Study

Davide Tosi(✉)[iD] and Marika Scalise

University of Insubria, via Ravasi 2, 21100 Varese, Italy
davide.tosi@uninsubria.it

Abstract. Software chess engines and advanced solutions based on Artificial Intelligence (AI) are changing the historical game of chess. Several AI-based chess engines are now available as closed or open software solutions with unique characteristics and game abilities. However, these AI-based engines are very complex to analyze from a functional point-of-view, and a deep study of their source code is not often enough to understand the advantages and disadvantages of their game approach. In this paper, we adopted a two-fold approach to study the behavior of a well-known AI-based chess engine called RubiChess. From one side, we studied the RubiChess architecture to understand its main game strategies and adopted algorithms. On the other side, we simulated a set of matches played against other AI chess engines, and we evaluated these matches (in different conditions) by using statistical tests and data visualization to determine the properties that made RubiChess unique. For example, the simulations highlight that RubiChess performs better as a white player and during "slow" games. This gives engine developers and players important insights into how RubiChess plays.

Keywords: Machine Learning · AI-based Chess Engines · NNUE · RubiChess · Data visualization

1 Introduction

The game of chess is a mentally challenging exercise with a fascinating history. The history of chess is not just about human victories; it is also a story of technological development. Chess engines and artificial intelligence programs have revolutionized the game, and the watershed moment in this evolution was the advent of Deep Blue [1], the chess engine that defeated the world champion, Garry Kasparov, in 1997. Deep Blue's success was not only due to its programming but also its innovative architecture, which showcased advancements in parallel processing and evaluation functions.

The evolution of these technologies has also changed the state of the art in chess, which refers to the new strategies, tactics, skills, and knowledge developed by experienced and competitive players over the years. Examining the latest and most innovative works in the field allows us to evaluate the quality and originality of games by comparing them with those of past and present masters. This analysis provides insights into the dynamic and ever-evolving landscape of chess intelligence and gameplay.

Since the creation of the first chess engine model, chess has been a field of experimentation and innovation in artificial intelligence. These engines have demonstrated extraordinary capabilities in calculating moves and analyzing chess positions, constantly pushing the limits of human knowledge. From those early beginnings to today, the evolution of chess engines has been astonishing, with hundreds of models developed, each bringing unique characteristics that distinguish them from others.

For this analysis, we will examine the architecture of RubiChess [5], a contemporary chess engine that is well-known for its balanced style of play. It implements advanced algorithms and heuristics to assess various positions on the board, as well as innovative techniques to analyze the game tree. By comparing the excellence and creativity of games with those created by experienced and current experts, we can better understand the continually changing and evolving chess intelligence and gameplay.

The main objectives of this paper are twofold:

1. deeply study the Artificial Intelligence (AI) engine of RubiChess to design a clear panoramic of the characteristics of this engine. This provides researchers with a precise picture of the behavioral AI model of the engine in the two types of roles of chess: as the white and black player;
2. compare the performance and behavioral AI model against other chess engines. This evaluation aimed to highlight the strengths and weaknesses of the engine. The comparison offers valuable insights into how RubiChess performs in different roles and situations, contributing to a deeper understanding of its strategic capabilities and potential areas for improvement.

The engine has been analyzed by studying the available documentation listed on the author's GitHub page. Additionally, extensive research has been conducted on the various properties of RubiChess, including its algorithms, the NNUE (Efficiently Updatable Neural Network) architecture, strategies, and skills. The analysis was conducted by simulating 16,824 games and collecting all necessary data.

The paper is organized as follows: Sect. 2 describes the workflow of RubiChess; Sect. 3 describes the methodology applied to evaluate the performance of the RubiChess against other chess engines; Sect. 4 provides a summary of our evaluation results, which help to satisfy the analysis and a discussion of the results obtained.

2 Design of the RubiChess Machine Learning Workflow

In this section, we analyze the main characteristics of the RubiChess engine and we sketch the workflow adopted by it to perform a chess match.

Chess engines based on Machine Learning (ML) abilities, such as AlphaZero, RubiChess or Stockfish, have a different type of approach in terms of playing chess despite the traditional chess engine [11]. The main purpose of traditional chess engines is to analyze various positions and select the optimal move based on that analysis. The engine evaluates each position by assigning a numerical value from the heuristic function, considering factors such as the balance of material, pawn structure, the safety of the king, the activity of the pieces and other relevant considerations.

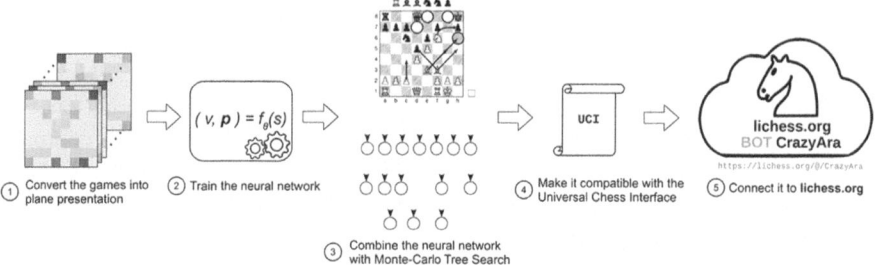

Fig. 1. Stockfish engine workflow

In the case of RubiChess, whose structure is similar to Stockfish, the procedure of playing the round is developed in several different steps (see Fig. 1 for a graphical representation of the workflow [12]):

1. Before calculating the most convenient move, it is relevant to have a representation of the current situation of the board; this can be achieved by assigning a numerical value or a vector to each square of the board to maintain chess positions for search, evaluation, and gameplay.
2. To evaluate the current position, chess engine as RubiChess, using the NNUE architecture. This architecture revolutionizes the chess engine technology by employing neural networks to assess game positions playing alpha-beta searching running on a CPU. Initially introduced into the Stockfish engine, now the NNUE architecture is adopted by numerous other chess engines. The NNUE architecture uses ML to produce a more efficient and accurate evaluation function. To do that, it applies its principle to matrix calculations inside a neural network [13]. Although evaluating a single position using a neural network may take longer instead of the traditional evaluation expression; it is possible to use the evaluation from the current position to speed up evaluations of very similar positions. This architecture can learn and recognize non-linear and complex relationships. It receives, stores, and uses experimental knowledge as stable states or maps in networks and recalls them when responding to the input variable.

The "neurons" have two functionality [17]:
   – summing the weighted inputs from different connections.
   – applying a transfer function to do that.

The values are transmitted to subsequent layers of neurons. The input layer receives the real-word (which is a strategy or a move) input and then advances it, in the form of weighted output, to the next layer as the figure above shows [14].

The output of the NNUE might be a single numerical, that represents the network's evaluation of the position's favorably for the current side or the other, or a set of scores for various features or properties of the position.

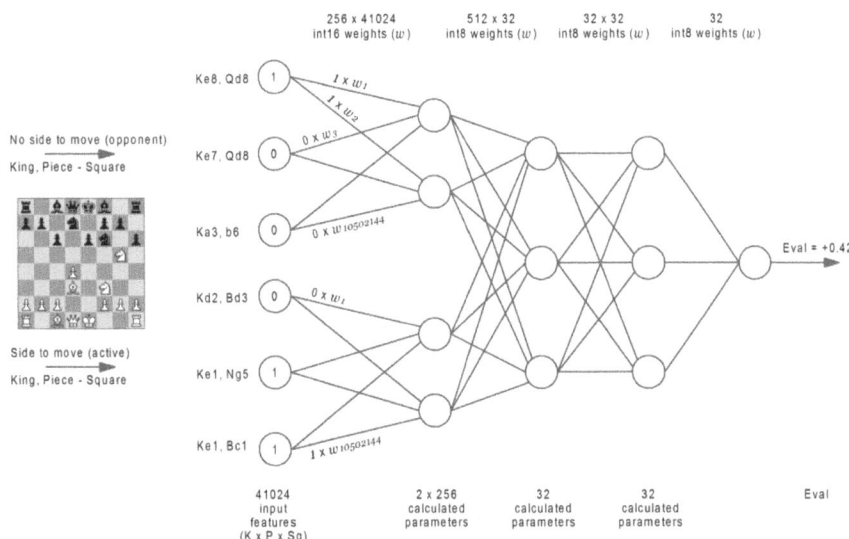

Based on the evaluations provided by NNUE, the engine makes use of a decision-making algorithm to determine the best move to play.

3. RubiChess employs a joining of traditional search algorithms, such as Alpha-Beta pruning, and the Principal Variation Search (PVS) technique to enhance its search capabilities [15]. The Alpha-Beta algorithm prunes branches of the game tree to focus on the most promising lines of play, while PVS explores the principal variation, which is the sequence of moves most likely to be played. This algorithm is a search algorithm that determines the best move for a player in a game tree. It does this by exploring all possible moves recursively, to maximize the value for the current player and minimize the value for their opponent.

This algorithm is efficient in reducing the number of explored nodes in the tree, saving the evaluation time.

The PVS algorithm takes as inputs:

Algorithm 1. Principal Variation Search (PVS) [3]

Require: Node *node*, Depth *depth*, α, β, Color *color*
Ensure: Best score α
1: **if** $depth = 0$ or *node* is a terminal node **then**
2:     **return** $color\times$ the heuristic value of *node*
3: **end if**
4: **for** each child of *node* **do**
5:     **if** child is the first child **then**
6:        $score \leftarrow -\text{pvs}(child, depth - 1, -\beta, -\alpha, -color)$
7:     **else**
8:        $score \leftarrow -\text{pvs}(child, depth - 1, -\alpha - 1, -\alpha, -color)$ {search with a null window}
9:        **if** $\alpha < \text{score} < \beta$ **then**
10:           $score \leftarrow -\text{pvs}(child, depth - 1, -\beta, -\alpha, -color)$ {if it failed high, do a full re-search}
11:        **end if**
12:     **end if**
13:     $\alpha \leftarrow \max(\alpha, \text{score})$
14:     **if** $\alpha \geq \beta$ **then**
15:        **break**   {beta cut-off}
16:     **end if**
17: **end for**
18: **return** α

- The current node of the search tree, where each node represents a possible move on the chessboard;
- The remaining depth of the search, which includes all the possible moves examined earlier based on the given game position;
- The best evaluation for the current player (represented by alpha);
- The best evaluation for the opposing player (represented by beta) and the color of the players, which can be either white or black.

For each child of the node, the following steps are performed:
(a) If it is the first child, a full recursive search is performed by calling the function with a **full search window** ($-\beta$ and $-\alpha$).
(b) For every other child, a recursive search is performed with a 'null' search window ($\alpha - 1$ and α).
(c) If the child's evaluation falls between α and β, a "fail-soft" or "Principal Variation Search" recursive search is performed.
(d) α is updated with the maximum between α and the child's evaluation.
(e) If α exceeds or equals β, a beta cut occurs and the search is aborted.

The output of the PVS search is the best evaluation score α

The decision of the move is based also on the strategy implemented into the engine which depends on the evaluations and the current state of the game. In computer chess, a strategy is a long-term plan that the engine uses to gain an advantage in the game. On the other hand, a tactic refers to a sequence of moves that the engine employs to gain an advantage over its opponent.

One of the strategies adopted by a chess player or chess motor, as in the case of RubiChess, is the implementation of the Opening Book [2], with the addition of the group of decisions and the planned moves that a player adopts to achieve the final aim, which could be the victory of the game or the achievement of an advantageous position.

Specifically, the Opening Book is an opening database which could be used as much as the opponent plays a new move from the database; usually, more common opening lines are stored at a higher depth than uncommon ones. As soon as the program is "out of book", it has to continue using normal search routines.

The opening and the well-studied collection of moves can influence the game's development. The chess opening is the sequence of the first moves done in a chess game. Knowing them can be essential for all the players. The chess motor, which uses the opening book, tends to obtain better performances than the others which don't use it. The usage of this book makes the chess motor more efficient in terms of decision speed and coherence at the beginning of a match, which can bring benefits to the game.

3 RubiChess Engine Evaluation

The detailed analysis of the RubiChess ML chess engine allows us to define a match simulation method to evaluate the advantages and disadvantages of the RubiChess engine.

In this section, we first describe the developed simulation method, and then we evaluate the RubiChess engine against other chess engines that use ML abilities.

The simulation method uses the BanksiaGui application [8], a "Live" chess GUI, and a Python script we developed to optimize the data acquisition process from match simulations.

With the BanksiaGui application, it is possible to simulate matches and see in "real-time" the score graph for each move to understand the performance of the engine.

The Python script allows the collection of data and specific information for each match, such as move time, ECO codes (Encyclopedia of Chess Openings), results obtained, and frequency of moves based on time decisions made by the engine. The script used the UCI (Universal Chess Interface) protocol to simulate chess games by allowing chess programs to communicate via the CPU.

```
game = chess.pgn.Game()
game.headers["Event"] = "*"
game.headers["Date"] = z.strftime(("%d %B %Y"))
game.headers["Time"] = z.strftime("%H:%M:%S")
game.headers["White"] = player_white
game.headers["Black"] = player_black
game.headers["TimeControl"] = f"{str(timeMove)}"
```

```
game.headers["PlayCount"] = str(play_count)
game.headers["Result"] = str(board.result())
game.headers["ECO"] = eco_.compare_(formatted_string, data)
```

The output of the simulations is saved in PGN (Portable Game Notation) files, from which relevant data are extracted into CSV (Comma-Separated Values) format. This information is the basis for conducting statistical tests and data visualization to evaluate the advantages and disadvantages of the RubiChess engine. It is important to note that the configuration of the simulation script was critical since we collected not only all the moves made during the match but also additional information, such as execution time and ECO ranking.

A PGN is a type of file that includes the data registered from chess matches. In this file are registered data, like the moves made by each of the two players, in algebraic notation that describes the type of materials moved and the coordinates of the destination square. A PGN file comprises three main subjects: properties, moves, and probable comments.

As an example, it follows an excerpt of a PGN file:

```
[Event "*"]
[Site "?"]
[Date "04 November 2023"]
[Round "ID"]
[White "Rubichess"]
[Black "Leela"]
[Result "1-0"]
[TimeControl "movetime: 1"]
[PlayCount "134"]
[ECO "C42g"]

1. e4 { 0.38 } 1... e5 { 0.38 }
2. Nf3 { -0.38 } 2... Nf6 { 0.45 }
3. ... ...
# 1-0
```

The excerpt of the PGN reports the name of the *event* in which the game was played (e.g., "World Chess Championship"), the *site* where the game was played, the game *date*, the ID associated with the *round*, the names of the *White* and *Black* players, the *Result* of the match as "1-0" (white victory), "0-1" (black victory) or "1/2-1/2" (draw), the *time control* to specify the maximum time the player has to make the next move, the incremental play count of the moves, the *ECO* representation of the opening used in the match, and then the sequence of moves done during the game with the relative score (positive or negative, assigned by Stockfish through a function which takes as input the situation of the chessboard and returns the relative value of the position) between curly brackets.

PGN files differ from each other because they collect different data for each time control (i.e., the time the engine can use to elaborate its next chess move).

In this paper, the time controls chosen for the different simulations are 0.1 s, 0.5 s, and 1 s. PGN files must be converted into CSV files to perform automatic analysis on the different simulations. To this end, we developed a Python script that parses the PGN file in two CSV files: the first CSV called "game_info" contains all the information of the different engines compared (such as the unique code to identify the game, the name of the white player, the name of the black player, the outcome of the game, the ECO code of the opening) while the second CSV is called "moves" and it contains all the information of the moves done (such as the notation, the move (c7c6, the pawn was moved from the square c7 to c6), the square where the piece stats and the square where the piece is arrived). The function used to convert the PGN to CSV files is PGNData from the library from converter.pgn_data.

Below are CSV's columns considered in our analysis:

The extract of the CSV relays the engines playing as *White player* and as *Black player*, the *result* of each game in the PGN file, the *ECO* code of the opening used, the *move* done by the current player, the square where the piece is moved and the square is moved, the *fen* (Forsyth-Edwards Notation) that is a standard notation used to describe the board position of a chess game; the *sequence of moves* done during a game.

Matches are then converted into graphical representations (e.g., bar charts, graphs, and pie charts) to understand and visually compare the trend of the most common moves used by each tested ML chess engine for each time control, the efficiency of the engine in terms of the frequency of the moves, the distribution of match's results, and the most common ECO used by RubiChess when playing as the white player to understand if there are some openings that are pre-arranged to employ in different time controls.

Moreover, we used different statistical tests to better understand the dependencies between the information listed in the PGN file. Specifically, we adopted:

- KS-test to do statistical tests between the games won by the RubiChess engine and all the results of the games included in the CSV file [9] (or the games won by the opponent engine considered) to understand if the two samples have the maximum cumulative difference without considered their distribution, and their standard deviations to understand if the data of the sample are constant or variable [16]. This test was used to determine the tendency of the matches based on the time control 0.5 and 1 sec. (e.g., matches played with time control 0.1 sec were not tested since the outcomes of them are mainly constant.) The function kstest from the scipy.stats Python library has been used.
- Chi-squared test [10] to figure out whether there is a statistical dependency between the performance of the match's results (i.e., victory, defeat, draw) and the time decision each engine had to make the move. A contingency table is used to divvy the outcomes collected in each time control, it was obtained from the function chi2_contigency from the scipy.stats Python library. The p-value is essential for determining which hypothesis to accept based on a significance level of 0.05.

Null Hypothesis is: the two categories are independent from each other, so the time control doesn't influence the result of the game.
Alternative Hypothesis is: the two categories are dependent on each other, so the time control influences the result of the game.

4 Results

This section firstly describes the performance of RubiChess through the comparison and visualization of the matches outcomes under different decision time conditions, and then summarizes the results of the statistical tests for RubiChess as the white and black player, respectively.

4.1 RubiChess as White Player

Analysis of Outcomes. As depicted in Fig. 2, the findings indicate that when the decision time is set at 0.1 s, the black player achieves a higher percentage of wins, suggesting that faster response times lead to greater efficiency (respectively, the white player wins 1944 matches, the black player 4055 matches). However, increasing the decision time to 0.5 s and 1 s, results in significant improvement in the white player's performance, with an increase in win percentage and a reduction in terminations other than victory (respectively, the white player wins 1284 and 187 matches, while the black player wins 262 and 45 matches). Although there has been an increase in the percentage of draws, this could imply that the competition between the engines is growing, resulting in more intriguing matches (respectively, 1214 matches in time 0.5 s and 295 matches in time one second). These results indicate that longer decision times could benefit the white player by allowing for a more thorough evaluation of moves and effective decision-making during the game.

Moreover, the study emphasizes the importance of considering decision time as a vital variable in evaluating the performance of chess engines in the context of artificial intelligence applied to the game of chess.

Analysis of the Results Obtained by the Tests. This assumption of the relation between the time decision and the outcome is confirmed by the result obtained from the Chi-squared test. The obtained results confirm our alternative hypothesis, which suggests a significant association between the variables considered, i.e., the time control and all possible match outcomes. The contingency table presents the number of cases observed for different possible outcomes and time controls used. The test results indicate significant differences between the collected and expected data, confirming that time control has a meaningful effect on the outcome of chess matches.

To be more specific, the Chi-square test calculates a value of 4493.67 with six degrees of freedom and a p-value $\ll 0.05$. This small value confirms strong statistical evidence against the null hypothesis (which suggests no association between the results and the time control used). The calculated values contradict

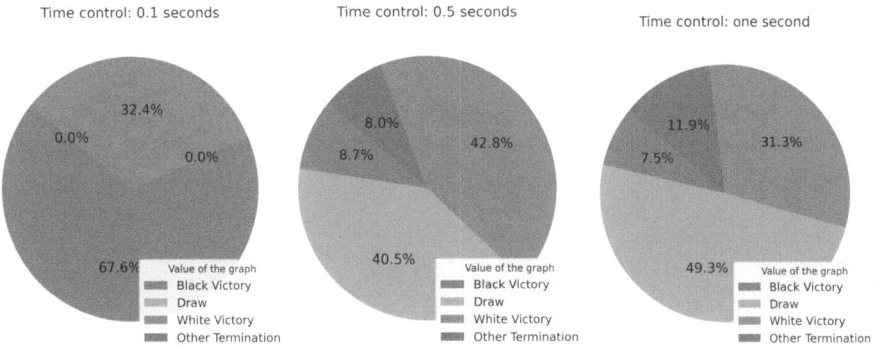

Fig. 2. Results distribution for matches conducted by RubiChess as White Player, with different conditions for time control

this notion and confirm the strong dependence between these two variables. Therefore, time control significantly influences the outcome of the game, as the significant discrepancy suggests between the observed and expected data.

The KS test shows a significant difference between the general distribution of matches played and different time controls. The KS value is over 0.5 for each time control (0.5 s and one second), indicating a considerable difference between the distributions. Additionally, the standard deviation of RubiChess is very low (rounded to 0.0) for each time control, suggesting that there is a consistency or minimum variance between the two-time decisions instead of the mean of the matches, so this low value indicates that the data points tend to vary less than their relative mean. The present means that the values are clustered around the mean of this variable.

These findings imply that RubiChess may employ unique game strategies or tactics that significantly impact the game's time. However, it is essential to consider that variability in playing times could affect other aspects of the chess engine's performance, such as the accuracy of moves or the overall strategy adopted.

4.2 RubiChess as Black Player

As in the white case, the study of the performance of Rubichess as the black player starts from the analysis of the results.

Analysis of Outcomes. The graphs of Fig. 3 show significant improvements in the performance of the black player with the increase of the time control. With time control of 0.1 s, the engine was not able to obtain victories against its opponents (overall 5968 matches), highlighting a weakness in its decision abilities that resulted in limited when it plays with restricted time that could be caused by the un-deepen search done during the exploration of the possible moves and the incomplete evaluation of them to find the best move.

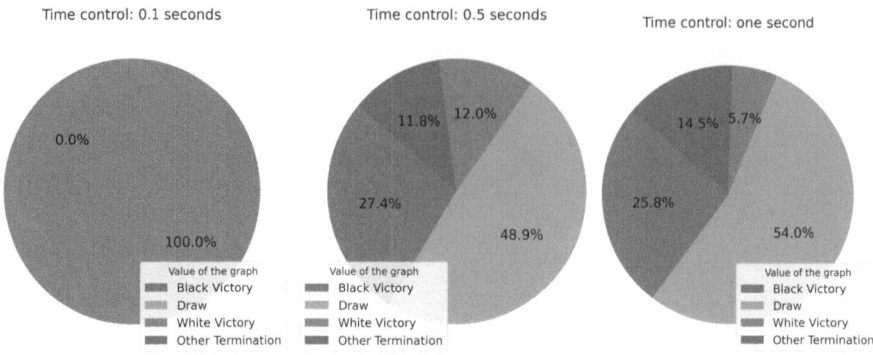

Fig. 3. Results distribution for matches conducted by RubiChess as Black Player, with different time controls

However, with a time control of 0.5 s, the engine has increased the percentage of victories (respectively, 575 matches was win), underlining better capacities to evaluate the moves with the time given. This could be possible not only because of the heightened search of the positions but also for the better identification of the opponent's weak or imprecise move to gain an advantage over it. Moreover, there is an important rise in the percentage of draws (respectively, 1027 matches) which could be due to the diversification of game lines and the balance between the engines, with a more accurate response to the opponent's attacks, using strategies more appropriate to equalize the match.

The percentage of draws increases more with the increase of the time control, i.e. one second, which contributes to the deeper search of the game variance with a more accurate evaluation of all the moves and strategies to adopt and the complexity and the competitiveness of the matches. There has been a growth in the balance between engine performance and tactical effectiveness (respectively, 109 matches over).

These findings reveal, as the case of the white player, a relevant relationship between time control, the RubiChess performance and the situation of the chess match.

Analysis of the Results Obtained by the Tests. The relation previously supposed between the time decision and the outcome has been confirmed by the Chi-squared test. This confirms our alternative hypothesis, which suggests a significant association between the variables considered.

The test results indicate significant differences between the collected and expected data, confirming that time control has a meaningful effect on the outcome of chess matches.

More specifically, the Chi-square test calculated has a value of 8743.76 with six degrees of freedom and a p-value $\ll 0.05$. This small value confirms strong statistical evidence against the null hypothesis (which suggests no association between the results and the time control used). The calculated values contradict this notion and confirm the strong dependence between these two variables.

The data suggests a significant difference between general match distribution and different time controls according to the KS test. The KS value exceeds 0.5 for both time controls, indicating a significant difference between the distributions.

This is supported by the standard deviation of RubiChess with a time control of 0.5 s, which evidences the distribution of the data tends to vary less than their relative mean, so they are clustered around the mean of this variable. However, the results of all the matches have a higher standard deviation compared to the Rubichess data, indicating that they are a little bit spread out.

This assumption could be confirmed by the results of the standard deviation considered (0.48 for RubiChess and 0.55 for all the outcomes), also for time control one second, which is similar to the previous. So, it's possible to affirm that RubiChess have the same distribution as the match outcomes, underlining flexibility over the situation of the game and its results reliable. This indicates a minimum difference between the time control and the dispersion that reveals uniform, clustered around the mean.

4.3 Results Discussion and Conclusion

RubiChess has the potential to become one of the most interesting modern engines based on the conducted analysis that reveals its flexibility to adapt to different games' situations. This engine can adapt to the characteristics of the opponent and can vary the opening based on the situation of the chessboard. Additionally, the engine carries out an Opening Book, which allows it to choose different types of openings based on the strategies it wants to employ.

There seems to be a flaw in RubiChess which causes it to repeatedly make the same moves when there isn't enough time to decide the best move to take advantage of the opponent. This flaw is evident in the graphs of the most frequently used moves, which remain the same regardless of the opponent. The reason for this could be the lack of time to analyze the game deeply enough. This issue is further confirmed by the graphs depicting RubiChess's performance against the opponents, where it loses all matches when the time control is set to 0.1.

It's essential to understand that the players' choices and strategies have a significant impact on the outcome of the game. Although some openings may statistically lead to faster or slower endgames, the actual flow of the game is entirely dependent on the specific moves and decisions made by the players.

Thanks to NNUE implementation, RubiChess can use supervised learning to improve its valuations of positions over time, learning from its defeats.

It is possible to observe a difference in performance between the white and black players in chess engines. The engines tend to perform differently based on the player they are playing as. When the engine plays as the white player and makes a quick decision, it tends to win more often and lose less frequently. Conversely, when playing as the black player, the engine tends to adopt a more competitive approach that increases with the amount of time set and evaluates moves based on the opponent's strategy.

Future work can involve a deep analysis about the most common moves used to understand and evaluate the strategies and tactics adopted by RubiChess when playing as white or black.

References

1. Campbell, M., Hoane, A.J., Jr., Hsu, F.: Deep blue. Artif. Intell. **134**(1), 57–83 (2002). https://doi.org/10.1016/S00043702(01)001291
2. Opening Book - Chessprogramming Wiki (2021). https://www.chessprogramming.org/
3. Wikiwand. Principal Variation Search (2021). https://www.wikiwand.com/. Accessed May 2024
4. Darius: Berserk. https://www.chessengeria.eu/berserk. Accessed 15 Jan 2024
5. Darius: RubiChess 20230407. Elo: 3595! (2023). https://www.chessengeria.eu/post/rubichess-20230407. Accessed May 2024
6. Darius: Koivisto 9.0 - released. https://www.chessengeria.eu/post/koivisto-9-0-released. Accessed 15 Jan 2024
7. Download Lc0 - Leela Chess Zero (2023). https://lczero.org/play/download/. Accessed May 2024
8. Hong Pham, N.: BanksiaGUI - A Free Chess Graphical User Interface (2019). https://banksiagui.com
9. Engmann, S., Cousineau, D.: Comparing distributions: the two-sample Anderson-Darling test as an alternative to the Kolmogorov-Smirnov test. J. Appl. Quant. Methods **6**, 1–17 (2011)
10. Conol, C., Peña, R.: Hypothesis Testing on Chess Openings (2020). https://cedricconol.github.io/chess-opening/
11. Panchal, H., Mishra, S., Shrivastava, V.: Chess moves prediction using deep learning neural networks. In: 2021 International Conference on Advances in Computing and Communications (ICACC), pp. 1–6 (2021). https://doi.org/10.1109/ICACC-202152719.2021.9708405
12. Stockfish 8 vs Gull 3 Free Chess Engine Match with free PGN download. https://copperbowl.de/3-Free-Chess-Engine-Match-with-free-PGN-download-94618.html. Accessed May 2024
13. Saumik. A Brief Guide to Stockfish NNUE. https://saumikn.com/. Accessed May 2024
14. Champion, A.: Dissecting Stockfish Part 2: In-Depth look at a chess engine (2022). https://towardsdatascience.com/search?q=stockfish
15. Isenberg, G.: RubiChess (2021). https://www.chessprogramming.org/RubiChess. Accessed May 2024
16. Elo rating system. https://en.wikipedia.org/wiki/Elo_rating_system. Accessed May 2024
17. Samadi, M., Azimifar, Z., Jahromi, M.Z.: Learning: an effective approach in endgame chess board evaluation. In: Sixth International Conference on Machine Learning and Applications (ICMLA 2007), pp. 464–469 (2007)

Evaluating Inductive Reasoning Capabilities of Large Language Models With The One Dimensional Abstract Reasoning Corpus

Cédric Mesnage[1,2](✉)[iD], Xiaoyang Wang[2][iD], Hang Dong[2][iD], and Aishwaryaprajna[2][iD]

[1] Institute for Data Science and Artificial Intelligence, University of Exeter, Exeter, UK
[2] Department of Computer Science, University of Exeter, Exeter, UK
{C.S.Mesnage,X.Wang7,H.Dong2,Aishwaryaprajna}@exeter.ac.uk

Abstract. We present an initial automated test to evaluate LLMs' capacity to perform inductive reasoning tasks. We use the GPT-3.5 and GPT-4 models to create a system which generates Python code as hypotheses for inductive reasoning to transform sequences of the One Dimensional Abstract Reasoning Corpus (1D-ARC) challenge. We experiment with three prompting techniques, namely standard prompting, Chain of Thought (CoT), and direct feedback. We provide results and an analysis of cost-to-success rate and benefit-cost ratio. Our best result is an overall 25% success rate with our CoT prompting on GPT-4, significantly surpassing the standard prompting approach. We discuss potential avenues to improve our experiments and test other strategies, and combine deductive reasoning with LLM-based inductive reasoning.

Keywords: Large Language Models · Inductive Reasoning · Prompt Engineering · Abstract Reasoning Corpus

1 Introduction

Pre-trained large language models (LLMs) are approaching or have exceeded human-level performance in various tasks including abstractive summarization, question answering in some domains, some coding tasks, etc. [2,14,15]. Moving towards more intelligent systems, or Artificial General Intelligence (AGI) [1], there are significant interests in the reasoning abilities of LLMs, which is a fundamental aspect of human intelligence [16]. There are debates about whether LLMs can reason based on the understanding of truth and logic, which is closer to the "thinking" process of humans [17]. By leveraging *prompting* techniques (in-context few-shot learning), pre-trained LLMs can solve some reasoning tasks in different benchmarks, including math, commonsense, and games [24,25], but their performance heavily depends on the prompting approach.

GPT-3 showcased the zero-shot inference ability, with limited ability to perform more complex reasoning tasks [26]. Few-shot learning is then studied to promote "thinking" by showcasing several examples of intermediate reasoning steps, named CoT [24]. Zero-shot CoT is proposed through simple additional prompting "*Let's think step-by-step*" [18]. While this research focuses on probing reasoning ability through in-context learning, we would like to further understand the variations of different prompting techniques. We would also like to understand the impact of feedback on the thinking process of LLMs, and how it would correct itself through feedback. Besides, there is a gap in benchmarking the cost of LLMs with different prompting techniques.

Inductive reasoning is a fundamental cognitive challenge that involves making generalizations from specific instances [27]. It requires inferring the principles from the observations and applying them to novel situations. Previous works on various types of reasoning tasks, including arithmetic, commonsense, and symbolic reasoning, have showcased the reasoning capabilities of LLMs to a certain degree [12,13,28,29]. Given that LLMs are in-context few-shot learners [26], these studies on reasoning abilities have paved the way for novel prompting and searching techniques, such as the CoT [24] and Tree of Thoughts [25].

Inspired by the Bayesian learner, a hypothesis search method is proposed to assist LLMs in solving complex inductive reason tasks [30]. Instead of directly prompting, LLMs are used to generate natural language hypotheses according to the problem description and examples, then translate hypotheses into Python programs for execution and testing. This work has demonstrated the contribution of hypothesis searching and computer programs in solving complex inductive reasoning problems. Note that human-annotated hypotheses are introduced as few-shot demonstrations for generating new hypotheses. Human annotators are also involved in selecting the subset of hypotheses, to avoid the huge computation load in testing all hypotheses.

Existing work demonstrates the improvement of LLMs' reasoning abilities through thinking step-by-step or introducing hypotheses selection and testing loops. We propose to perform a comprehensive analysis of different prompting techniques, as well as different ways for LLMs to produce results, with minimal human intervention in the reasoning process. As the study [30] has demonstrated the contribution of computer programs in solving complex inductive reasoning problems, we also ask LLMs to generate computer programs for easy generalisation and testing. Thus, this task is relevant to program synthesis, a classical problem studied with inductive and deductive reasoning and more recently with neural networks and LLMs [22,23]. We mainly test LLMs for inductive reasoning in this work and will discuss deductive reasoning at the end of the paper.

We are looking at evaluating the effect of CoT [24] and CoT variations [18] on solving reasoning tasks from the 1D-ARC [5] which we compare with standard prompting and direct feedback. We use the OpenAI API and store every interaction with it for each solving method that we set. The goal of developing intelligent agents which would be capable of understanding the world by interpreting their observations requires theoretically formalising a model of the real world. Along this goal, we argue that observing a phenomenon, hypothesising and theorising about it requires the same skills as finding the transformation rules in the 1D-ARC tasks and that code is a way of formalising a theory. We evaluate the inductive reasoning capacity of LLMs as intelligent agents by automatically generating prompts to perform 1D-ARC tasks, prompting for code of transformation functions and testing the correctness of the generated functions.

The remainder of the paper is as follows: Sect. 2 describes the 1D-ARC, prompting techniques, testing and evaluation metrics; Sect. 3 summarises the results; and Sect. 4 and Sect. 5 discuss the results and conclude the work with potential future research.

2 Methodology

This section describes the methods we developed to evaluate and compare different prompting methods. In order to evaluate the level of reasoning of current large language models, we develop a method to automatically assess the correctness of answers produced by an LLM. We choose the 1D-ARC as a dataset of tasks to perform since it provides us with a clear test to evaluate.

2.1 Abstraction and Reasoning Corpus

The 1D-ARC consists of 900 tasks grouped into 18 categories of 50 tasks each. For each task, we have 3 examples of input-to-output sequences to illustrate the transformation to be performed and one test set of one input and one output. Figure 1 is a graphical representation of each category. These are tasks which are simple to complete for humans. The sequences are represented as colored squares, the background ones are black and the colors of interest as other colors. For instance, a transformation can be to shift all colored squares by one pixel to the right whilst keeping the black pixels identical. Another transformation is the mirror, which transforms the sequence into its reverse.

To evaluate the capacity of LLMs to understand those transformations, we generate prompts sent to the OpenAI application programming interface (API) for GPT-3.5/4 models for each of the tasks and ask for Python code to be generated. We describe this process in the following subsections.

2.2 Code Generation Process

Firstly we encode the sequences as strings of integers, 0 representing black and other digits for other colors. We prompt an LLM to produce a "transform (sequence)" Python function which returns the transformed sequence. The Python function is generated given the few transformation examples provided in the prompt.

2.3 Prompt Engineering

We experimented on a trial-and-error basis by interacting with the GPT 3.5/4 models directly in order to engineer a prompt which produces a usable function

Fig. 1. The 1D-ARC task categories and representative input-output pairs [5].

without producing too much unnecessary text. In fact, the GPT models tend to add many comments in their code even for very simple tasks which increases the produced tokens and therefore the cost. The OpenAI API for GPT models requires a "system" input, in which we describe the purpose of the query and what we expect as an output and a "user" entry in which we give the transformation examples. The API returns a response as JSON which includes the answer to our query, the number of input and output tokens and multiple choices depending on how many we asked for. We built three different prompting techniques, namely standard prompting, CoT and direct feedback which we describe next and provide comparative results.

Standard Prompting. Our first method is standard prompting, namely prompting the LLM solely for a function solving the task. Figure 2 is an example of the generated prompt for task 1dMove2p7, the response from GPT-4 and the test we perform. No help is given to the LLM, clue or hint, solely the sequences of the transformations examples.

Since this is the less costly of our methods, for each of the tasks we prompt both GPT-3.5-turbo and GPT-4 for 10 choices using standard prompting.

Chain of Thought. The CoT concept is to prompt the LLM to produce reasoning steps or decompose a task or question before giving an answer, since those produced tokens become the context, it is thought that it increases the accuracy of the answer given and has been used in various contexts.

There are multiple ways to perform CoT. One could give an example of the expected reasoning process, prompting for a "step-by-step" list, for reasoning in a controlled or free-form manner. As shown in Fig. 3, we chose to add to our standard prompting method the prompt: "Provide your reasoning step by step prior to giving the function.".

The investigated LLMs do produce a description of the transformation and of the task to complete before writing the Python function. We will see in the results section how this affects the success rate as well as the cost of running the experiment for all 900 tasks.

Direct Feedback. Seeing that many of the functions produced either do not run or fail at completing the tasks, we designed a different method called direct feedback in which we iteratively query GPT models, letting it know of previous functions on the same task which did not pass the test. Figure 4 shows an example of the prompt we generate on a second iteration once a function has failed. When a function succeeds at the task we stop iterating. We also stop if we reach the maximum number of iterations, in our experiment is 5.

The functions which failed the tasks are given in the prompt following the sentence: "Knowing that these functions failed:".

Direct feedback being iterative is much more costly than standard prompting, especially when the size of the input increases with the number of functions that failed previously (the number of iterations).

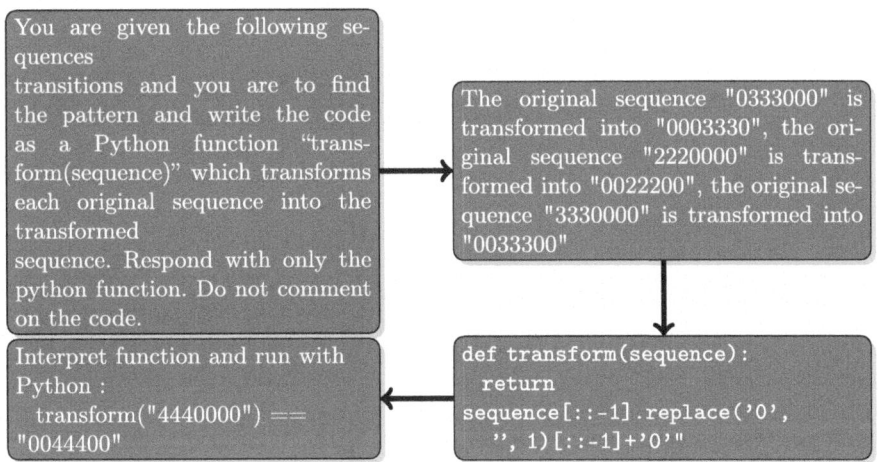

Fig. 2. Standard prompting of the task 1dMove2p7, the answer from GPT-4 and the test we run.

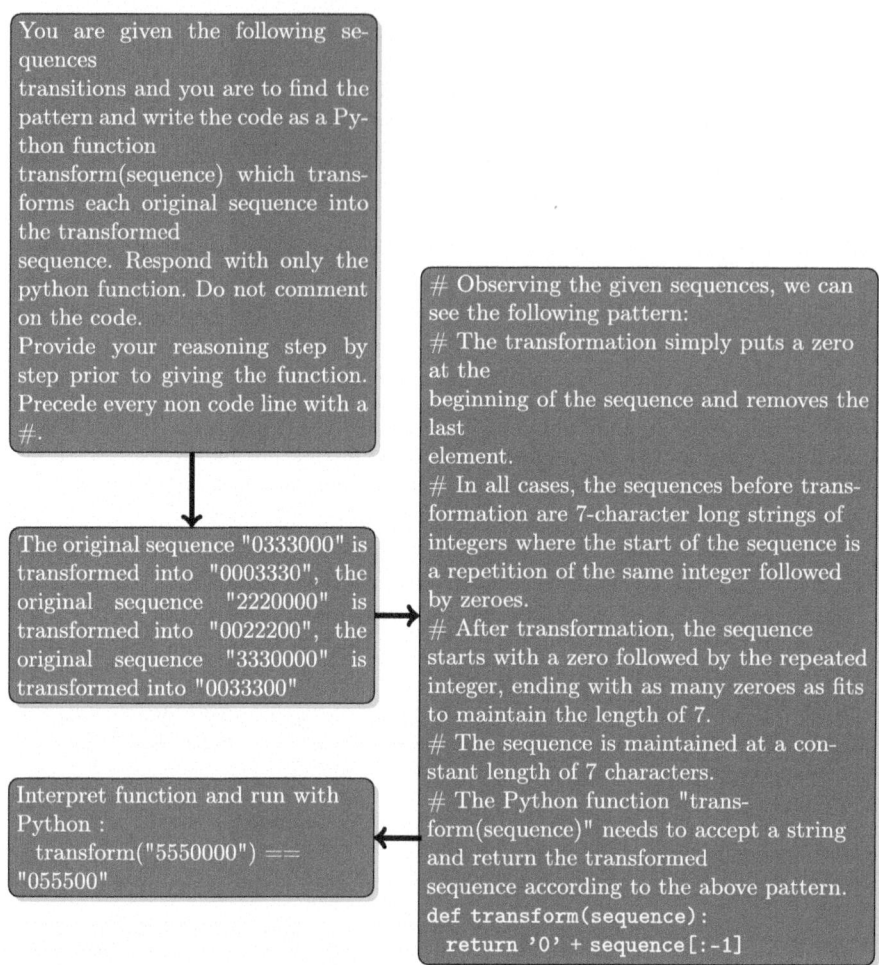

Fig. 3. Generated CoT prompt for the task 1dMove1p44, the response from GPT-4 and the test we run.

2.4 Testing Generated Code

To test the functions responded by the GPT models we first process the response, removing any additional lines that would not be coded and removing comments. The code is then interpreted and if it can be interpreted will be accessible as a function. We then run this function to test if the transformation is correct by comparing the returned sequence with the output sequence from the dataset. We have guardrails when running the code, for instance, some functions run into infinite loops which we kill after 1 s. We store all responses from GPT-3.5/4, results, and calculations on Zenodo[1] for reference.

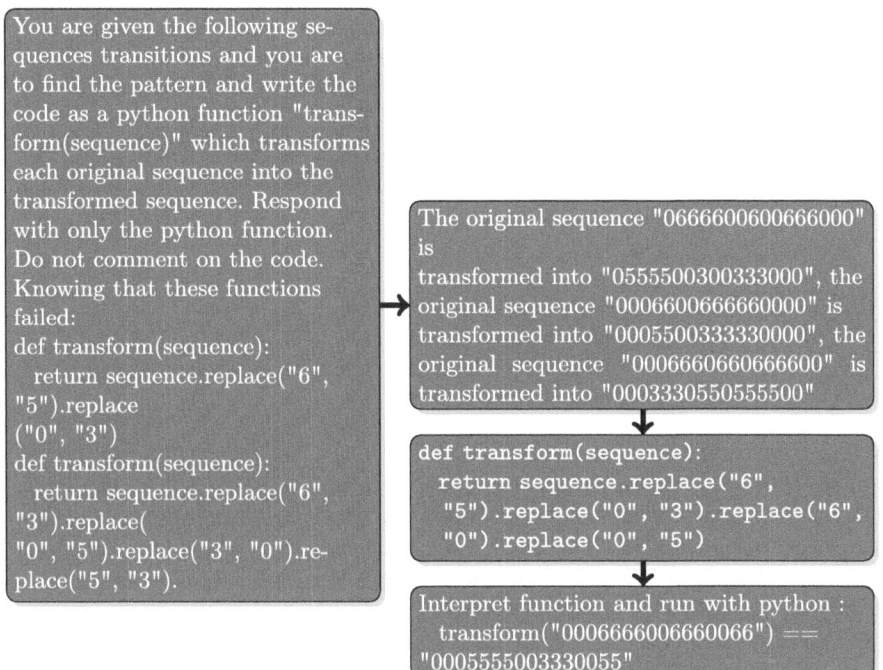

Fig. 4. Direct feedback iteration for the task 1dMove2p7, the answer from GPT-4 and the test we run.

2.5 OpenAI API Parameters

When querying the chat completion API we can specify parameters which affect the response generation. We adjust the number of choices to generate, when an LLM generates a completion, it selects the most likely next token according to a probability distribution on all tokens, OpenAI provides multiple choices resulting from choosing within those distributions. We set the temperature to 1 for all experiments. We did not specify a maximum number of tokens.

2.6 Evaluation Metrics

To evaluate and compare our different prompting strategies, we compute metrics on the data resulting from testing, i.e. the number of successful tasks per task category and per number of choices.

---

[1] Our implementation is available at https://zenodo.org/records/13735926. The repository also contains the 1D-ARC dataset from https://github.com/khalil-research/1D-ARC [5].

$$\mu = \sum_{t \in \text{tasks}} \frac{\text{success}(t)}{|\text{tasks}|} \tag{1}$$

We define the mean success rate μ as the sum of successful tasks divided by the total number of tasks. The cost c per method and number of choices is defined in USD by OpenAI as input_tokens $* 0.0000005 +$ output_tokens $* 0.0000015$ with the GPT-3.5-turbo model and input_tokens $* 0.00003 +$ output_tokens $* 0.00006$ with GPT-4.

We define the benefit-cost ratio as the mean success rate divided by the cost. We calculate it for each prompting strategy and number of choices/iterations.

$$\text{bcr} = \frac{\mu}{c} \tag{2}$$

The following section analyses the results of our experiments.

3 Results

We have queried the OpenAI API with both the GPT-3.5-turbo and GPT-4 models. The tables summarising the results are given in the appendix for clarity. With either model CoT performs better for the same number of choices but standard prompting reaches a similar success rate whilst remaining less costly. Even with GPT-4 and CoT, the LLM did not find a solution for any task of several task categories, namely Pattern Copy, Pattern Copy Multicolor and Mirror which do not seem as such complex tasks for a human, the most successful tasks are not necessarily the simplest, since the LLM is more successful at the Move by 2 pixels tasks than by 1 pixel, we think that there is a large amount of chance in solving the task hence why there is no clear rate per complexity of task correlation emerging from the results which would reveal some inductive reasoning from the LLM.

3.1 GPT-3.5-Turbo Results

Tables 1, 2, 3 give the number of successful tasks out of 50 tasks per category. Tables 4, 5, 6 give the number of tokens used for input and output per amount of choices as well as the calculation of the mean success rate, the cost and benefit-cost ratio. The best μ is with 5 choices. The CoT strategy has a 2.44% success rate, which is really low, but slightly better than 10 choices with the standard prompting approach, which has a 1.56% success rate. The best bcr is with the CoT strategy and 2 choices, the bcr drops with more than 2 choices queried.

Figure 5a and 5b represent visually the results. The success rates and costs discussed in Fig. 5a show tradeoffs for all prompting approaches. However, the slope of the tradeoff for the direct feedback approach is less steep than the rest of the approaches. This means that even if the cost increases significantly, the increase in success rate is not that significant for the direct feedback approach.

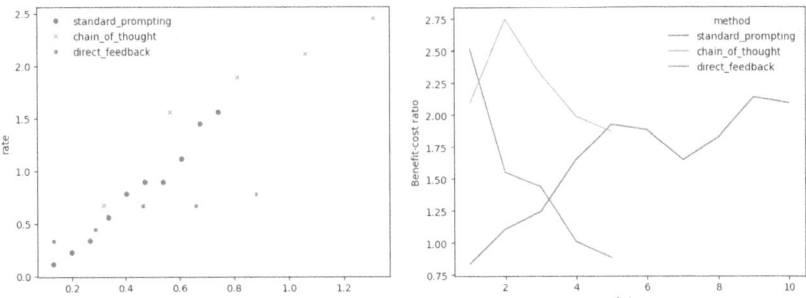

(a) Tradeoff analysis between success rate and cost.

(b) Benefit-cost ratio per number of choices.

Fig. 5. Analysis of ChatGPT-3.5-turbo results.

3.2 GPT-4

Tables 7, 8, 9 give the number of successful tasks per category, out of 50 tasks per category. Tables 10, 11, 12 give the number of tokens used for input and output per amount of choices as well as the calculation of the mean success rate, the cost and benefit-cost ratio.

Figure 6a and 6b represent the results visually. Contrarily with the GPT-3.5 model, the best *bcr* is with 1 iteration of direct feedback whereas the *bcr* of standard prompting maximises at 4 choices. The CoT success rate is lower and decreases with the number of choices. The best success rate obtained is with 5 choices and the CoT strategy with 25.11% whereas standard prompting with 10 choices reaches 19.77%. However, the standard prompting is generally much less costly.

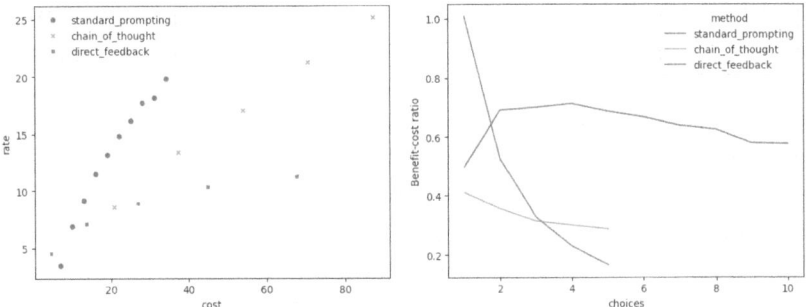

(a) Tradeoff analysis between success rate and cost.

(b) Benefit-cost ratio per number of choices.

Fig. 6. Analysis of GPT-4 results.

Figure 6a shows that the tradeoff slope for standard prompting is the steepest, followed by the CoT and direct feedback prompting approaches. As in the case of the GPT-3.5 model, in the GPT-4 model as well, a significant increase in the cost does not guarantee a significant increase in the success rates with the direct feedback approach.

4 Discussion

We observe generally low results with the LLMs, GPT-4 and GPT-3.5-turbo. GPT-4 obtained the best successful rate of around 25%, although greatly better (about 10 times) than the results compared to GPT-3.5-turbo. This suggests that the greater parameter size enhanced their capability with inductive reasoning with the prompts. While the results are still low, the performance of GPT-4 is encouraging. This shows the potential of prompts (i.e., forward propagation) of very large neural networks of billions of artificial neurons to approximate inductive reasoning. Further ways of learning with the prompts, e.g., instruction tuning and reinforcement learning, may help improve the performance.

The generally low results are also likely due to our numeric representation (e.g. "000333000") of the 1D-ARC task, given that the tokenisation process may split the numeric string (e.g., by splitting it into "000", "333", "000", separately), and thus may distort the meaning of the original string. Future studies can improve the numeric representation when prompting the model, or using other open LLMs (e.g., the Llama series [21]) which have a distinct processing of numeric strings.

We have run the standard prompting experiment with a Python list of integer embedding of the sequences, such as in [0,0,0,3,3,3,0,0,0] instead of "000333000" to compare and in fact there is an improvement, the success rate at 10 choices is 5% compared to 1.55% with GPT-3.5, 45 tasks out of 900 for a cost of $1.11 as opposed to $0.74. This result remains low and therefore shows the limited inductive reasoning capability of LLMs.

It might be possible to increase the performance of LLMs at coding for inductive reasoning by developing prompting techniques and we have shown that zero-shot CoT does improve the success rate. Nevertheless, the transformer architecture of LLMs and the autoregressive approach does not enable effective inductive reasoning and research in this direction is necessary.

5 Conclusion and Future Work

In this work, we have explored the capability of off-the-shelf LLMs, especially GPT-3.5 and GPT-4 for inductive reasoning using the 1D-ARC corpus. Results show great room for improvement for future systems that employ LLM. Program synthesis has been used as an intermediary task to solve the problem. While we mainly explored pure inductive reasoning with LLMs, the area of program synthesis has been explored greatly with both inductive and deductive methods.

These methods can be explored in the future to generate programmable hypotheses to solve the tasks from the Abstract Reasoning Corpus.

Finally, the low performance of a pure LLM-based approach in this work may suggest the need for future studies to combine inductive and deductive methods with large neural networks like LLMs. For example, Retrieval Augmented Generative (RAG) [20] together with symbolic representation in deductive reasoning (e.g., graph-based RAG) and other LLM adaptation methods like fine-tuning with reinforcement learning may provide new avenues to combine neural-based inductive and deductive methods.

Appendix

Table 1. Standard prompting number of successful functions per task category out of 50 by the number of choices with GPT-3.5-turbo. The categories of Padded Fill, Move 2 Towards, Flip, Mirror, Denoise, Denoise Multicolor, Pattern Copy, Pattern Copy Multicolor, Recolor by Odd Even, Recolor by Size, Recolor by Size Comparison, and Scaling were omitted as none of the tasks were successful.

Choice	1	2	3	4	5	6	7	8	9	10
Task category										
Move 1	1	1	1	2	2	2	2	2	2	3
Move 2	0	0	0	0	0	0	0	1	2	2
Move 3	0	1	2	2	3	3	3	4	4	4
Move Dynamic	0	0	0	0	0	1	1	1	2	2
Fill	0	0	0	1	2	2	2	2	2	2
Hollow	0	0	0	0	0	0	0	0	1	1
Mean success rate	**0.11**	**0.22**	**0.33**	**0.56**	**0.78**	**0.89**	**0.89**	**1.11**	**1.44**	**1.56**
Standard deviation	**0.47**	**0.64**	**1.02**	**1.33**	**1.83**	**1.84**	**1.84**	**2.19**	**2.35**	**2.52**

Table 2. Direct feedback number of successful functions per task category out of 50 by the number of choices with GPT-3.5-turbo. The categories of Move 2, Move 2 Towards, Move Dynamic, Fill, Padded Fill, Flip, Mirror, Denoise, Denoise Multicolor, Pattern Copy, Pattern Copy Multicolor, Recolor by Odd Even, Recolor by Size, Recolor by Size Comparison, and Scaling were omitted since none of those tasks were successful. Mean success rate is a percentage.

Choice	1	2	3	4	5
Task category					
Move 1	0	1	2	2	2
Move 3	2	2	2	2	3
Hollow	1	1	2	2	2
Mean success rate	**0.33**	**0.44**	**0.67**	**0.67**	**0.78**
Standard deviation	**1.029**	**1.097**	**1.534**	**1.534**	**1.833**

Table 3. CoT number of successful functions per task category out of 50 by the number of choices with GPT-3.5-turbo. The categories of Pattern Copy, Mirror, Pattern Copy Multicolor, Recolor by Odd Even, Recolor by Size, Recolor by Size Comparison, and Scaling were omitted since none of those tasks were successful. Mean success rate is a percentage.

Choice	1	2	3	4	5
Task category					
Move 1	3	5	6	6	6
Move 2	0	1	1	1	1
Move 3	0	1	1	1	1
Move Dynamic	1	1	1	1	1
Move 2 Towards	0	0	0	0	1
Fill	1	2	2	2	2
Padded Fill	0	0	0	0	0
Hollow	1	2	3	5	7
Flip	0	1	1	1	1
Denoise	0	1	1	1	1
Denoise Multicolor	0	0	1	1	1
Mean success rate	**0.67**	**1.56**	**1.89**	**2.11**	**2.44**
Standard deviation	**1.534**	**2.526**	**3.027**	**3.462**	**4.033**

Table 4. Standard prompting used tokens, success rate and costs per number of choices with GPT-3.5-turbo. Mean success rate is a percentage.

Number of choices	1	2	3	4	5
Input tokens	129607	129607	129607	129607	129607
Output tokens	45289	90578	135867	181156	226446
Mean success rate	0.111	0.222	0.333	0.556	0.778
Cost in $	0.133	0.201	0.269	0.337	0.404
Benefit-cost ratio	0.837	1.107	1.241	1.651	1.923
Number of choices	6	7	8	9	10
Input tokens	129607	129607	129607	129607	129607
Output tokens	271735	317024	362313	407602	452891
Mean success rate	0.889	0.889	1.111	1.444	1.556
Cost in $	0.472	0.54	0.608	0.676	0.744
Benefit-cost ratio	1.882	1.645	1.827	2.136	2.09

Table 5. CoT used tokens, success rate and costs per number of choices with GPT-3.5-turbo. Mean success rate is a percentage.

Number of choices	1	2	3	4	5
Input tokens	143122	143122	143122	143122	143122
Output tokens	165003	330007	495010	660014	825017
Mean success rate	0.667	1.556	1.889	2.111	2.444
Cost in $	0.319	0.567	0.814	1.062	1.309
Benefit-cost ratio	2.089	2.746	2.32	1.989	1.867

Table 6. Direct feedback used tokens, success rate and costs per number of choices with GPT-3.5-turbo. Mean success rate is a percentage.

Number of choices	1	2	3	4	5
Input tokens	129964	308737	528124	789198	1092075
Output tokens	45080	88221	132960	177407	223007
Mean success rate	0.333	0.444	0.667	0.667	0.778
Cost in $	0.133	0.287	0.464	0.661	0.881
Benefit-cost ratio	2.514	1.55	1.438	1.009	0.883

Table 7. Standard prompting number of successful functions per task category out of 50 by the number of choices with GPT-4. The categories of Pattern Copy, Pattern Copy Multicolor, Mirror, Recolor by Odd Even, Recolor by Size, and Recolor by Size Comparison were omitted since none of those tasks were successful. Mean success rate is a percentage.

Choice	1	2	3	4	5	6	7	8	9	10
Task category										
Move 1	7	11	13	14	17	19	20	21	21	25
Move 2	2	4	4	5	5	5	5	7	8	9
Move 3	2	8	12	13	13	14	16	18	18	18
Move Dynamic	2	2	2	3	4	4	4	6	6	6
Move 2 Towards	0	0	0	0	0	1	1	2	2	3
Fill	5	13	19	22	22	24	26	27	28	30
Padded Fill	0	0	0	0	0	1	1	1	1	1
Hollow	6	10	14	18	22	26	29	30	30	32
Flip	5	7	9	13	19	21	22	24	24	27
Denoise	0	4	4	5	5	5	6	6	6	7
Denoise Multicolor	1	1	1	4	5	6	8	10	12	13
Scaling	1	2	4	6	6	7	7	7	7	7
Mean success rate	3.44	6.89	9.11	11.44	13.11	14.78	16.11	17.67	18.11	19.78
Standard deviation	4.74	8.81	12.17	14.25	16.38	18.28	19.88	20.81	21.04	22.93

Table 8. CoT number of successful functions per task category out of 50 by the number of choices with GPT-4. The categories of Pattern Copy, Pattern Copy Multicolor, and Mirror were omitted since none of those tasks were successful. Mean success rate is a percentage.

Choice	1	2	3	4	5
Task category					
Move 1	7	10	12	14	18
Move 2	5	15	17	22	27
Move 3	6	8	10	15	19
Move Dynamic	2	2	2	4	6
Move 2 Towards	1	2	3	3	4
Fill	5	11	14	17	23
Padded Fill	1	1	1	1	1
Hollow	16	25	28	37	39
Flip	9	11	20	21	24
Denoise	5	8	12	13	16
Denoise Multicolor	15	21	24	31	34
Recolor by Size	1	2	2	2	2
Recolor by Size Comparison	0	0	1	1	1
Scaling	4	4	7	10	12
Mean success rate	**8.56**	**13.33**	**17.0**	**21.22**	**25.11**
Standard deviation	**9.889**	**15.262**	**18.153**	**22.949**	**25.743**

Table 9. Direct feedback number of successful functions per task category out of 50 by the number of choices with GPT-4. The categories of Mirror, Pattern Copy, Pattern Copy Multicolor, Padded Fill, Recolor by Odd Even, Recolor by Size, and Recolor by Size Comparison were omitted since none of those tasks were successful. Mean success rate is a percentage. Mean success rate is a percentage.

Choice	1	2	3	4	5
Task category					
Move 1	5	10	13	14	17
Move 2	1	2	4	4	4
Move 3	4	5	5	6	6
Move Dynamic	1	1	1	3	3
Move 2 Towards	1	1	2	3	3
Fill	9	13	15	15	16
Hollow	10	15	20	23	23
Flip	5	8	9	11	11
Denoise	1	1	2	4	5
Denoise Multicolor	3	4	5	6	6
Scaling	1	4	4	4	7
Mean success rate	**4.56**	**7.11**	**8.89**	**10.33**	**11.22**
Standard deviation	**6.28**	**9.634**	**11.985**	**13.092**	**13.791**

Table 10. Standard prompting used tokens, success rate and costs per number of choices with GPT-4. Mean success rate is a percentage.

Number of choices	1	2	3	4	5
Input tokens	129607	129607	129607	129607	129607
Output tokens	50767	101533	152300	203067	253834
Mean success rate	3.444	6.889	9.111	11.444	13.111
Cost in $	6.934	9.98	13.026	16.072	19.118
Benefit-cost ratio	0.497	0.69	0.699	0.712	0.686
Number of choices	6	7	8	9	10
Input tokens	129607	129607	129607	129607	129607
Output tokens	304600	355367	406134	456900	507667
Mean success rate	14.778	16.111	17.667	18.111	19.778
Cost in $	22.164	25.21	28.256	31.302	34.348
Benefit-cost ratio	0.667	0.639	0.625	0.579	0.576

Table 11. CoT used tokens, success rate and costs per number of choices with GPT-4. Mean success rate is a percentage.

Number of choices	1	2	3	4	5
Input tokens	143122	143122	143122	143122	143122
Output tokens	276249	552498	828748	1104997	1381246
Mean success rate	8.556	13.333	17.0	21.222	25.111
Cost in $	20.869	37.444	54.019	70.593	87.168
Benefit-cost ratio	0.41	0.356	0.315	0.301	0.288

Table 12. Direct feedback used tokens, success rate and costs per number of choices with GPT-4. Mean success rate is a percentage.

Number of choices	1	2	3	4	5
Input tokens	129607	309496	541517	828076	1168956
Output tokens	10624	71260	180380	335254	542357
Mean success rate	4.556	7.111	8.889	10.333	11.222
Cost in $	4.526	13.56	27.068	44.957	67.61
Benefit-cost ratio	1.007	0.524	0.328	0.23	0.166

References

1. Morris, M.R., et al.: Levels of AGI: Operationalizing Progress on the Path to AGI. arXiv preprint: arXiv:2311.02462 (2023)
2. OpenAI: GPT-4 Technical Report. arXiv preprint: arXiv:2303.08774 (2023)

3. Xu, B., Ren, Q.: Artificial open world for evaluating AGI: a conceptual design. In: Goertzel, B., Iklé, M., Potapov, A., Ponomaryov, D. (eds.) Artificial General Intelligence. AGI 2022. LNCS, vol. 13539, pp. 452–463. Springer, Cham (2023). https://doi.org/10.1007/978-3-031-19907-3_43
4. Hernández-Orallo, J., Loe, B.S., Cheke, L., Martínez-Plumed, F., Héigeartaigh, S.Ó.: General intelligence disentangled via a generality metric for natural and artificial intelligence. Sci. Reports **11**(1), 22822 (2021)
5. Legg, S.: Machine super intelligence. Ph.D. dissertation, Università della Svizzera italiana (2008)
6. Torrado, R.R., Bontrager, P., Togelius, J., Liu, J., Perez-Liebana, D.: Deep Reinforcement Learning for General Video Game AI. In: 2018 IEEE Conference on Computational Intelligence and Games (CIG), IEEE (2018)
7. Schrittwieser, J., et al.: Mastering Atari, go, chess and shogi by planning with a learned model. Nature **588**(7839), 604–609 (2020)
8. Taylor, R., et al.: GALACTICA: a large language model for science. In: Proceedings of the Artificial General Intelligence Conference (2022)
9. Burda, Y., et al.: Large-scale study of curiosity-driven learning. arXiv preprint arXiv:1808.04355 (2018)
10. Sutton, R.S., Barto, A.G.: Reinforcement Learning: an Introduction. MIT Press (2018)
11. Bakker, P.B.: The State of Mind Reinforcement Learning with Recurrent Neural Networks. Ph.D. thesis (2004)
12. Zhao, Z., Lee, W.S., Hsu, D.: Large language models as commonsense knowledge for large-scale task planning. In: Advances in Neural Information Processing Systems, vol. 36 (2024)
13. Wang, S., Wei, Z., Choi, Y., Ren, X.: Can LLMs Reason with Rules? Logic Scaffolding for Stress-Testing and Improving LLMs. arXiv preprint arXiv:2402.11442 (2024)
14. Zhang, T., Ladhak, F., Durmus, E., Liang, P., McKeown, K., Hashimoto, T.B.: Benchmarking large language models for news summarization. In: Transactions of the Association for Computational Linguistics, vol. 12, pp. 39–57. MIT Press (2024)
15. Bommarito II, M., Katz, D.M.: GPT takes the bar exam. arXiv preprint arXiv:2212.14402 (2022)
16. Häggström, O.: Artificial general intelligence and the common sense argument. In: Müller, V.C. (eds) Philosophy and Theory of Artificial Intelligence 2021. PTAI 2021. Studies in Applied Philosophy, Epistemology and Rational Ethics, vol. 63, pp. 155–160. Springer, Cham (2021). https://doi.org/10.1007/978-3-031-09153-7_12
17. Wang, B., Yue, X., Sun, H.: Can ChatGPT defend its belief in truth? Evaluating LLM reasoning via debate. In: Findings of the Association for Computational Linguistics: EMNLP 2023, pp. 11865–11881 (2023)
18. Kojima, T., Shane Gu, S., Reid, M., Matsuo, Y., Iwasawa, Y.: Large language models are zero-shot reasoners. In: Advances in Neural Information Processing Systems, vol. 35, pp. 22199–22213 (2022)
19. Potapov, A., et al.: General-purpose minecraft agents and hybrid AGI. In: Goertzel, B., Iklé, M., Potapov, A., Ponomaryov, D. (eds) Artificial General Intelligence. AGI 2022. LNCS, vol. 13539, pp. 75–85. Springer, Cham (2023). https://doi.org/10.1007/978-3-031-19907-3_8
20. Lewis, P., et al.: Retrieval-augmented generation for knowledge-intensive NLP tasks. In: Advances in Neural Information Processing Systems, vol. 33, pp. 9459–9474 (2020)

21. Touvron, H., et al.: Llama: Open and efficient foundation language models. arXiv preprint arXiv:2302.13971 (2023)
22. Liu, J., Xia, C.S., Wang, Y., Zhang, L.: Is your code generated by ChatGPT really correct? rigorous evaluation of large language models for code generation. In: Advances in Neural Information Processing Systems, vol. 36, pp. 21558–21572 (2023)
23. Kalyan, A., Mohta, A., Polozov, O., Batra, D., Jain, P., Gulwani, S.: Neural-guided deductive search for real-time program synthesis from examples. In: International Conference on Learning Representations, pp. 1–15 (2018)
24. Wei, J., et al.: Chain-of-thought prompting elicits reasoning in large language models. In: Advances in Neural Information Processing Systems, vol. 35, pp. 24824–24837 (2022)
25. Yao, S., et al.: Tree of thoughts: deliberate problem solving with large language models. arXiv preprint, arXiv:2305.10601 (2023)
26. Brown, T., et al.: Language models are few-shot learners. Adv. Neural. Inf. Process. Syst. **33**, 1877–1901 (2020)
27. Han, S.J., Ransom, K.J., Perfors, A., Kemp, C.: Inductive reasoning in humans and large language models. Cognitive Systems Research, vol. 83, p. 101155 (2024)
28. Imani, S., Du, L., Shrivastava, H.: Mathprompter: Mathematical reasoning using large language models (2023). arXiv:2303.05398
29. Wang, K., et al.: Mathcoder: Seamless code integration in LLMs for enhanced mathematical reasoning. arXiv:2303.05398 (2023)
30. Wang, R., Zelikman, E., Poesia, G., Pu, Y., Haber, N., Goodman, N.D.: Hypothesis Search: Inductive Reasoning with Language Models. arXiv preprint, arXiv:2309.05660 (2023)

Program Synthesis Using Inductive Logic Programming for the Abstraction and Reasoning Corpus

Filipe Marinho Rocha[1,3](✉), Inês Dutra[1,3], Vítor Santos Costa[1,3], and Luís Paulo Reis[2,4]

[1] Faculdade de Ciências da Universidade do Porto, Rua do Campo Alegre, 4169-007 Porto, Portugal
up199803145@up.pt
[2] Faculdade de Engenharia da Universidade do Porto, Rua Dr. Roberto Frias, 4200-465 Porto, Portugal
[3] INESCTEC, Campus da FEUP, Rua Dr. Roberto Frias, 4200-465 Porto, Portugal
[4] LIACC, FEUP, Rua Dr. Roberto Frias, 4200-465 Porto, Portugal

Abstract. The Abstraction and Reasoning Corpus (ARC) is a general artificial intelligence benchmark that is currently unsolvable by any Machine Learning method. It demands strong generalization and reasoning capabilities, which are known to be weaknesses of Neural Network-based systems. In this work, we propose a Program synthesis system to solve ARC, Induce Logic Programs for Abstract Reasoning (ILPAR), which casts an ARC problem as a sequence of Inductive Logic Programming (ILP) problems. We have implemented a simple Domain Specific Language (DSL) that corresponds to a small set of object-centric abstractions relevant to ARC. This is Background Knowledge used by ILP to create abstract Logic Programs that provide reasoning capabilities to our system. When solving each ARC task, ILPAR can generalize from a few training examples: pairs of Input-Output grids. The Logic Programs are able to generate objects present in the Output grid and the combination of these can transform an Input grid into an Output grid. We randomly chose some tasks from ARC that do not require more than the small number of primitives we implemented in our DSL and showed that providing only this to ILPAR, it can solve tasks that require each different reasoning.

Keywords: Program synthesis · ILP · Logic Programming · ARC

1 Introduction

Machine Learning [25], more specifically, Deep Learning (DL) [20], has achieved great successes and even surpassed human performance in several fields [33]. However, these successes are in what is called narrow Artificial Intelligence (AI), since each DL model is prepared to solve a specific task very well, but fails to solve different kinds of tasks [16, 19].

It is known Artificial Neural Networks (ANNs) and DL lack generalization capabilities. Their performance is greatly reduced when applied to Out-of-Distribution data [10–12].

More recently, Large Language Models (LLMs) [31] have shown impressive abilities, shortening the gap between Machine and Human Intelligence. However, they still show a lack of reasoning capabilities and require a lot of data and computation [32,34,36].

The ARC challenge [1] was designed by François Chollet in order to evaluate whether AI systems have progressed at emulating human-like general intelligence. It was presented in 2019, but it remains an unsolved challenge, and current DL models, such as LLMs, cannot solve it [2,4]. GPT-4V, which is GPT4 enhanced for visual tasks [18], is also unable to solve it [5,7,17].

ARC is composed of a Training set with 400 tasks and an Evaluation set with 400 tasks. There is also a Test set with 100 tasks, which is completely private. All of these tasks require very different reasoning to solve. As Chollet describes, ARC requires Developer-aware generalization, which is a stronger form of generalization than Out-of-Distribution generalization [1].

For a researcher setting out to solve ARC, it is perhaps best understood as a benchmark for Program synthesis [1]. Program synthesis [23,24] is a subfield of AI with the purpose of generating programs that satisfy a high-level specification, often provided in the form of example pairs of inputs and outputs for the program, which is exactly the ARC format, as can be seen in Fig. 1.

Chollet recommends starting with developing a domain-specific language (DSL) capable of expressing all possible solution programs for any ARC task. Since the exact set of ARC tasks could be anything that involves Core Knowledge priors [1], this is a challenge. Object awareness is considered one of the Core Knowledge prior of humans, necessary to solve ARC tasks [1,8,9] and is central to general human visual understanding [13]. There is previous work using Object-centric approaches to ARC, that shows its usefulness [6,9].

Inductive Logic Programming (ILP) [22] is a Machine Learning method, known for being able to learn and generalize from few training examples [26,27] that can perform Program synthesis [21]. To the best of our knowledge, it was never applied to the ARC challenge.

We present a Program synthesis system called ILPAR, that uses ILP, on top of a small set of Object-centric abstractions, manually defined. It does Program synthesis by searching for relations between objects in the task examples and then describing these by Logic Programs. The full program is composed by a sequence of Logic Programs that are capable of generating objects in an empty Output Grid, in order to reach the solution.

We selected five random tasks that contain only simple objects and relations between these that our small DSL can represent, and applied our system to solve them.

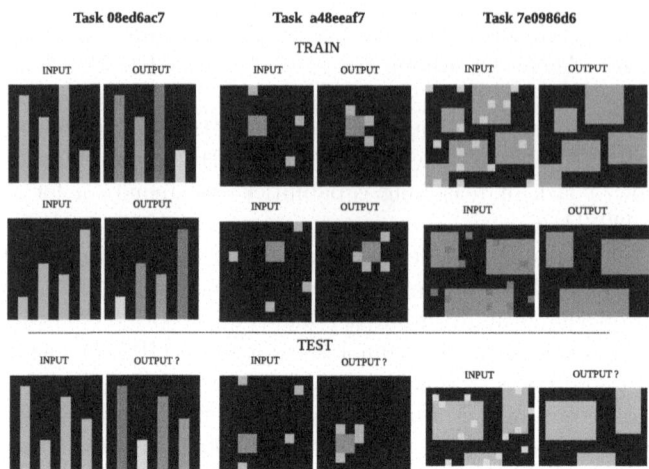

Fig. 1. Example tasks of the ARC Training dataset with the solutions shown. The goal is to produce the Test Output grid given the Test Input grid and the Train Input-Output grid pairs. We can see the logic behind each task is very different and the relations between objects are key to the solutions.

2 Object-Centric Abstractions and Representations

Object-centric abstractions substantially reduce the search space by enabling the focus on the relations between objects, instead of individual pixels or irrelevant pieces of the grids. However, there may be multiple ways to interpret the same image in terms of objects; therefore, we keep multiple, overlapping Object representations for the same image. Likewise, there may be multiple relations that explain an image transformation that may intersect in some space of the Output image, that is, they generate objects in the Output grid that may overlap, so we keep multiple Relation representations as well.

2.1 Objects and Relations Between Objects

We have manually defined a simple DSL composed of the objects: Point, Line, and Rectangle, defined in Table 1, and relations between objects: LineFromPoint, Translate, Copy, PointStraightPathTo, defined in Table 2.

Table 1. Object types

Objects & Attributes
Point & (x, y, color)
Line & (x1, y1, x2, y2, color, len, orientation, direction)
Rectangle & (x1, y1, x2, y2, x3, y3, x4, y4, color, clean, area)

Table 2. Relations types

Relations & Attributes
LineFromPoint & (point, line, len, orientation, direction)
Translate & (obj1, obj2, xdir, ydir, colorobj2)
Copy & (obj1, obj2, colorobj2, cleanobj2)
PointStraightPathTo & (point, obj, xdir, ydir, orientation, direction)

2.2 Multiple Representations

An image representation in our Object-centric approach is defined by a list of objects and a background color. This representation can build a target image from an empty grid by filling the grid with the background color and then drawing each object of the list on top of it.

An image grid can be defined by multiple Object representations. For example, a single Rectangle in an empty grid can also be defined as several contiguous Line or Point objects. Since we do not know the logic behind the transformation of the Input grid into the Output grid, we do not know which Object representations will be more relevant to the solution.

In Fig. 2, we can see an example of, depending on the Input-Output transformation logic required, one particular Object description is more appropriate than the other. In this case, all transformations may require only one relation: Copy, and only vary on the Object selected as the Input: a Point, Line, or Rectangle.

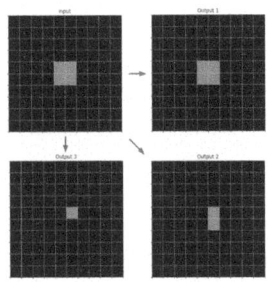

Fig. 2. Example of an Input grid with a Rectangle and three possible Output grids depending if Copy is applied to the full Rectangle or just a Line or a Point inside it.

Likewise, the same image transformation can be explained by different relations. Looking at Fig. 2, if we want to explain the Input-Output 1 transformation, we have Copy applied to the rectangle, which is the simplest explanation. But, LineFromPoint applied to the two top points in the rectangle, with length two and direction Down, would build the same Rectangle in the Output.

Since ARC tasks are made up of multiple training examples, the simplest explanation for an individual example could be different from the one common to multiple examples. So, we work with multiple and intermingled representations of objects and relations until we get the final program or programs that can reach the solution. Only then, we can know what Object and Relation representations were essential to the solution.

Likewise, at a higher level, different combinations of Logic Programs applied separately can produce the same Input-Output grids transformation, but at the end, we select the shortest program following the Occam's razor principle. So, we also keep multiple program representations until we reach the end of the search for a solution.

3 ILP

Inductive Logic Programming (ILP) is a form of logic-based Machine Learning. The goal is to induce a hypothesis, in the form of a Logic Program or set of Logical Rules, that generalizes given training examples and Background Knowledge (BK). Our DSL, composed of objects and relations, constitutes a part of the BK.

As with other forms of ML, the goal is to induce a hypothesis that generalizes the training examples. However, while most forms of ML use tensors to represent data, ILP uses logic programs. Unlike most forms of ML that learn functions, ILP learns relations.

Sometimes, reaching a logical definition of a complex relation requires obtaining a very large program, that is, one composed of many rules or many terms in the body of the rule(s). Learning a large program with ILP is very challenging, but there are some approaches that try to overcome this, such as Divide-and-conquer [29].

The fundamental ILP problem is to efficiently search a large hypothesis space in order to build a Logic Program. There are ILP approaches that search in either a top-down or bottom-up fashion, and others that combine both.

3.1 Divide-and-Conquer

Divide-and-conquer approaches in ILP, divide the examples into subsets, and search for a program for each subset. We use a Divide-and-conquer approach, but instead of applying it to examples, we divide each individual example into subsets of objects present in the example, like in previous work on ARC [14].

Instead of searching for a Logic Program that transforms an Input image into an Output image, we search for a Logic Program that can generate a subset of the Output image, that is, can generate one or more Objects present in the Output image.

3.2 Top-Down

In a Top-down approach, one starts from the most general Horn clause, that covers all the examples: positive and negative, iteratively adding literals to its body, to cover more positive and less negative examples, as long as it does not become overly specific. In a Bottom-up approach, one would start from a long clause that may cover just one example, iteratively removing literals from it, until it becomes overly general.

We use a top-down approach to clause building, inspired by FOIL [30].

4 Program Synthesis Using ILP

ILP is usually seen as a method for Concept Learning but is capable of building a Logic Program in Prolog [28], which is Turing-complete, hence it performs Program synthesis. We extend this, by combining Logic Programs to build a bigger program that generates objects in sequence, to fill an empty grid, which corresponds to, procedurally, applying grid transformations to reach the solution, as shown in Fig. 6.

A Relation can be used to generate objects. All of the relations in our DSL, except for PointStraightPathTo, are able to generate Objects given the first Object of the Relation. To generate a particular output, the Relation should be constrained by a Logic Program.

Example of a logic program in Prolog that defines the application of LineFromPoint:

```
target(point(X,Y,C),Line,Len,Orientation,Direction):-
   point(X,Y,C),
   equal(Len,5),
   equal(Orientation,'vertical'),
   line_from_point(point(X,Y,C),Line,Len,Orientation,Direction).
```

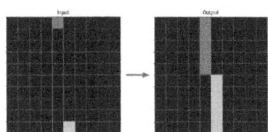

Fig. 3. Example of an Input-Output transformation when applied a LineFromPoint Logic Program that is able to generate Lines from Points.

The literal: point(X,Y,C), in the rule, is grounded with facts that were added to the BK, including the objects present in the Input Grid: the red and blue points, shown in Fig. 3. This Logic Program can generate unambiguously two lines in the Output. The variable Direction is not needed for this, since the points are on the edges of the Grid and so, can only grow into lines in one Direction each: down and up, given the Orientation: vertical. If we would remove the literal that instantiates the variable Orientation from the body of the rule, it would generate multiple Lines from each Input Point, as shown in Fig. 4.

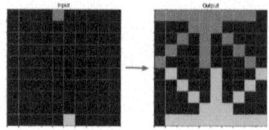

Fig. 4. Example of an Input-Output transformation when applied a LineFromPoint Logic Program with a defined Length but without having the Orientation defined.

So, by obtaining Logic Programs through ILP, we are indeed constructing a program that generates objects that can fill an empty Test Output grid, in order to reach the solution.

5 System Overview

5.1 Objects and Relations Retrieval

Our system begins by retrieving all Objects defined in our DSL, present in the Input and Output grids, for each example of a task.

Then we search for the relations defined in our DSL, that hold between the found objects. We search for Relations between Objects present in the Input Grid: Input-Input Relations, present in the Output Grid: Output-Output Relations and between Objects in the Input Grid and Output Grid: Input-Output Relations.

The types of objects previously found can constrain the search for relations, since some relations are specific to some kind of objects. We save all of the grounded objects and relations retrieved as facts in the BK to be used by ILP.

5.2 ILP Calls

We then call ILP to create Logic Programs to define only the relations found in the previous step, which reduces the space of the search.

We start with the Input-Output Relations, since we need to build the Output objects using information from the Input. After we generate some Object(s) in the Output grid, we can also start using Output-Output relations to generate Output Objects from other Output Objects.

Input-Input relations can exist in the body of the rules but would not be the target Relations to be constrained by ILP, since they do not generate any Object in the Output, but still, they can provide useful information, in order to do so.

The grounded relations and objects returned by the ILP call are added to the BK and considered now as if they were present in the initial Input data.

Besides updating the BK, we keep the Logic Programs being created and the updated versions of the Output Grid states after applying these Logic Programs. The first ILP call receives as input, empty Output grid states, but then with each subsequent ILP call, the updated Output Grid states are taken as the input, where to build upon.

Candidate Generation. The candidate literals to be added to the body of the Rule or Horn Clause in the ILP process, can be formed by the predicates that correspond to the Objects and Relations defined in Tables 1 and 2.

We provide additional predicates to form candidates: equal/2, greater Than/2, lowerThan/2, member/2. greaterThan and lowerThan, relate only numeric variables. We also provide linearCombination/4, which consist of $Y=aX+b$ formulations as candidate terms, X and Y being numeric variables and a and b, numeric constants in some interval we pre-define. We finally also provide grid/2 which contain the variables X and Y dimensions of a grid.

Since we are using typed objects and relations with typed attributes, we only generate those candidates that relate each of the variables of the Relation to variables or constants of the same type.

For example, in building a Logic Program that constrains a LineFromPoint relation, we are going to generate candidates that instantiate Points or attributes of Points. Then, we proceed to the other variables besides Line that remain free, because it is the object we want the relation to generate. Len is of type Integer, so it is only relevant to be related to other Int-typed variables or constants.

Positive and Negative Examples. As we have seen in Sect. 3, the usual approach to clause building requires Positive and Negative examples. However, the ARC dataset is made up only of positive examples. So we created a way to represent negative examples.

In our system for each Logic Program being constructed, the Positive examples are the Objects that this program generates that exist in the Training Data, and the Negative examples are the Objects that the program generates but do not exist in the Training Data.

For example, consider that the Output in Fig. 3 is the correct one, but the program generates all the lines in the output of Fig. 4, like the shorter clause we mentioned. For this program, the positives it covers would be the two lines in Fig. 3 and the negatives covered would be all the lines in Fig. 4, except for the vertical ones, which are eight lines.

Here, our Object-centric approach also reduces the complexity of the problem, compared with taking the entire Grid into account, to define what is a Positive example or not.

Matching Between Training Examples. A single ILP call is made for a target Relation taking into account all of the Task's Train examples. Our ILP system is constrained to produce programs that match between examples, that is, the same program works in the same way when applied to different examples.

These Logic Programs should be abstract enough and generalize to two or more of the training examples. Why not all the training examples? One of the five example tasks we selected showed us that finding a program that solves all Training examples may be too complex and unnecessary.

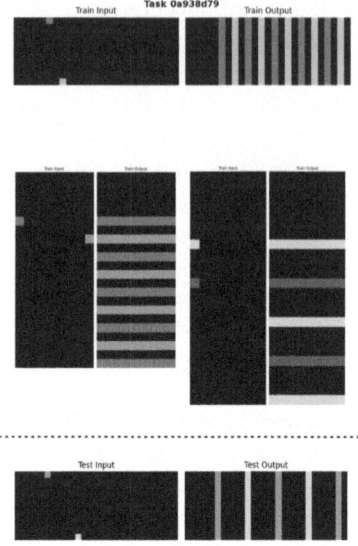

Fig. 5. Example task that contains two Train examples with vertical lines and two Train examples with horizontal lines, in the Output grids. The Test example requires vertical Lines in the same way as the first two Train examples.

In Fig. 5, we can see a sample task and identify the steps for its solution: draw Lines from Points until the opposite border of the grid and then Translate these Lines repeatedly in the perpendicular direction of the lines until the end of the grid. To find the program that executes these steps would be more complex than finding a solution that only matches the first two examples, but would also solve the Test example.

The Test example only needs two translations in the Output Grid, but a final program with more translations would work, since it would fill the entire grid and the extra translations would just not apply. Our system considers this to be a valid program.

But if the Test grid was longer and required more translations than present in the Train examples, our program would produce an incomplete solution. For this kind of tasks, we would need to use higher-order constructs, to apply the same target Relation program, a specific number of times or until some condition fails or is triggered.

5.3 Rules, Grid States and Search

The first ILP call is made on empty Output grid states. Each ILP call produces a rule that defines the way a Relation generates objects and returns new Output grid states with these objects added, for each example. Each subsequent ILP call is made on the updated Output Grid states.

The search is based on finding the right sequence of Logic Programs that can build the Output Grids, starting from the empty Output grid, like in Fig. 6. When we have a program that builds at least two complete Train Output Grids and can also produce a valid solution in the Test Output Grid, we consider it a final program.

A valid program is one that builds a consistent theory. For example, a program that generates different colored lines that intersect with each other is inconsistent, since both Lines cannot exist in the Output grid. We consider programs like this invalid and discard them in our search.

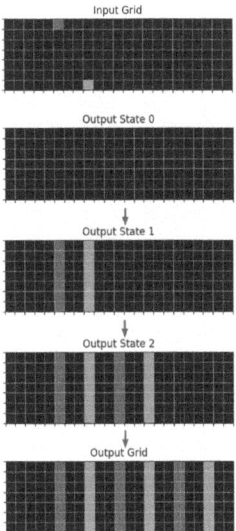

Fig. 6. Example of a successful sequence of transformations to the Output grid state reaching the final state which is the correct Output grid in one of the ARC task Train examples.

5.4 Deductive Search

In Fig. 7, we can see an example Task that in order to be solved, it would be easier to do it in reverse, that is, the Input Grid being generated from the Output grid, but since we cannot solve the Test example in this way, we use a form of deductive search to overcome this.

When we apply a Logic Program constructed by ILP, we can have different outputs depending on the order of the Object generation. When an object is generated in the Output grid, no other object generated afterwards can intersect with it.

So, the coverage of grid space of the same Logic Program may vary, depending on the order of the program application. It is the procedural aspect of our system. When applying several Logic Programs in sequence, this problem gets even bigger.

So, we save all the ways that each of the Logic Programs being created can generate objects and the combinations of these, when joining the Logic Programs into the full program. We choose the combinations that maximize space coverage in the Training Output Grids.

When applying the full program to produce the Test Output grid, we use a deductive search to find the way of applying it that covers the most surface. Since the final program can reproduce the whole surface of Train Output grid(s), we should have a solution that can reproduce all of the Test Output grid too.

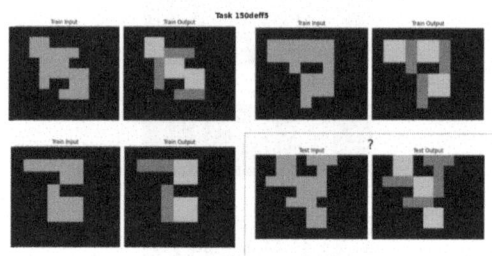

Fig. 7. Example task where the Output Grids are more informative than the Input Grids, to define and apply the object Relations for the solution.

6 Experiments

Our system was successfully applied to five tasks: the three tasks in Fig. 1, the task in Fig. 5, and the task in Fig. 7. In the appendices, we present the Output solutions for each task in Prolog.

7 Conclusion

ILPAR was able to solve the five selected sample tasks. When we finish its software implementation, we will apply our system to the full dataset and gather more insights about our approach.

ILP is at the core of our system and we showed that by providing it with only a small DSL as BK, it is able to learn from few Train examples and construct and represent the diverse logic behind solutions of ARC tasks. ILP gives our system abstract learning and generalization capabilities, which is at the core of the ARC challenge.

Since the other ARC tasks may depend on many other DSL primitives, we plan to develop a way to automate DSL creation. As we mentioned earlier, we are going to need higher-order constructs to solve other potential tasks, and we also plan to incorporate these into our system.

Acknowledgement. This work is financed by National Funds through the Portuguese funding agency, FCT -Fundação para a Ciência e a Tecnologia, within project UIDB/50014/2020.

A Tasks Solutions

A.1

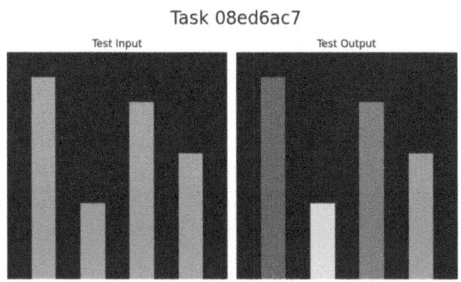

Task 08ed6ac7

```
target(line(X11,Y11,X12,Y12,Color1,Len1,Orientation1,Direction1),Line,Color_out,Clean):-
    line(X11,Y11,X12,Y12,Color1,Len1,Orientation1,Direction1),
    line(X21,Y21,X22,Y22,Color2,Len2,Orientation2,Direction2),
    line(X31,Y31,X32,Y32,Color3,Len3,Orientation3,Direction3),
    line(X41,Y41,X42,Y42,Color4,Len4,Orientation4,Direction4),
    equal(Color_out,'blue'),
    equal(Clean,100),
    greaterthan(Len1,Len2),
    greaterthan(Len1,Len3),
    greaterthan(Len1,Len4),
    copy(line(X11,Y11,X12,Y12,Color1,Len1,Orientation1,Direction1),Line,Color_out,Clean).

target(line(X11,Y11,X12,Y12,Color1,Len1,Orientation1,Direction1),Line,Color_out,Clean):-
    line(X11,Y11,X12,Y12,Color1,Len1,Orientation1,Direction1),
    line(X21,Y21,X22,Y22,Color2,Len2,Orientation2,Direction2),
    line(X31,Y31,X32,Y32,Color3,Len3,Orientation3,Direction3),
    line(X41,Y41,X42,Y42,Color4,Len4,Orientation4,Direction4),
    equal(Color_out,'red'),
    equal(Clean,100),
    greaterthan(Len1,Len2),
    greaterthan(Len1,Len3),
    lowerthan(Len1,Len4),
    copy(line(X11,Y11,X12,Y12,Color1,Len1,Orientation1,Direction1),Line,Color_out,Clean).

target(line(X11,Y11,X12,Y12,Color1,Len1,Orientation1,Direction1),Line,Color_out,Clean):-
    line(X11,Y11,X12,Y12,Color1,Len1,Orientation1,Direction1),
    line(X21,Y21,X22,Y22,Color2,Len2,Orientation2,Direction2),
    line(X31,Y31,X32,Y32,Color3,Len3,Orientation3,Direction3),
    line(X41,Y41,X42,Y42,Color4,Len4,Orientation4,Direction4),
    equal(Color_out,'green'),
    equal(Clean,100),
    greaterthan(Len1,Len2),
    lowerthan(Len1,Len3),
    lowerthan(Len1,Len4),
    copy(line(X11,Y11,X12,Y12,Color1,Len1,Orientation1,Direction1),Line,Color_out,Clean).
```

```
target(line(X11,Y11,X12,Y12,Color1,Len1,Orientation1,Direction1),Line,Color_out,Clean):-
    line(X11,Y11,X12,Y12,Color1,Len1,Orientation1,Direction1),
    line(X21,Y21,X22,Y22,Color2,Len2,Orientation2,Direction2),
    line(X31,Y31,X32,Y32,Color3,Len3,Orientation3,Direction3),
    line(X41,Y41,X42,Y42,Color4,Len4,Orientation4,Direction4),
    equal(Color_out,'yellow'),
    equal(Clean,100),
    lowerthan(Len1,Len2),
    lowerthan(Len1,Len3),
    lowerthan(Len1,Len4),
    copy(line(X11,Y11,X12,Y12,Color1,Len1,Orientation1,Direction1),Line,Color_out,Clean).
```

A.2

Task a48eeaf7

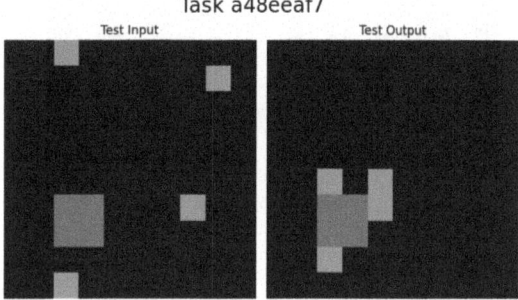

```
target(point(X,Y,C),Point,X_dir,Y_dir,Color_out):-
    point(X,Y,C),
    rectangle(X1,Y1,X2,Y2,X3,Y3,X4,Y4,Color,Clean,Area),
    point_straight_path_to(
        point(X,Y,C),
        rectangle(X1,Y1,X2,Y2,X3,Y3,X4,Y4,Color,Clean,Area),
        X_dir1,
        Y_dir1,
        Orientation,
        Direction),
    equal(X_dir,X_dir1),
    equal(Y_dir,Y_dir1),
    equal(Color_out,C),
    translate(point(X,Y,C),Point,X_dir,Y_dir,Color_out).

target(rectangle(X1,Y1,X2,Y2,X3,Y3,X4,Y4,Color,Clean,Area),Rectangle,Color_out,Clean_out):-
    rectangle(X1,Y1,X2,Y2,X3,Y3,X4,Y4,Color,Clean,Area),
    equal(Color_out,Color),
    equal(Clean_out,Clean),
    copy(
        rectangle(X1,Y1,X2,Y2,X3,Y3,X4,Y4,Color,Clean,Area),
        Rectangle,
        Color_out,
        Clean_out).
```

A.3

Task 7e0986d6

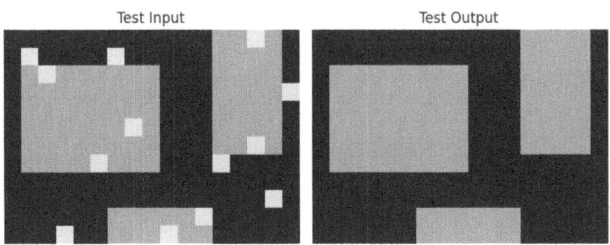

```
target(rectangle(X1,Y1,X2,Y2,X3,Y3,X4,Y4,Color,Clean,Area),Rectangle,Color_out,Clean_out):-
    rectangle(X1,Y1,X2,Y2,X3,Y3,X4,Y4,Color,Clean,Area),
    equal(Color_out,Color),
    equal(Clean_out,100),
    copy(
        rectangle(X1,Y1,X2,Y2,X3,Y3,X4,Y4,Color,Clean,Area),
        Rectangle,
        Color_out,
        Clean_out).
```

A.4

Task 0a938d79

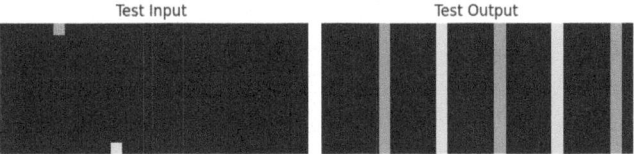

```
target(point(X,Y,C),Line,Len,Orientation,Direction):-
    grid(X_dim,Y_dim),
    point(X,Y,C),
    equal(Len,X_dim),
    equal(Orientation,'vertical'),
    line_from_point(point(X,Y,C),Line,Len,Orientation,Direction).

target(line(X1,Y1,X2,Y2,Color,Len,Orientation,Direction),Line,X_dir,Y_dir):-
    point(X,Y,C),
    point(X11,Y11,C11),
    line(X1,Y1,X2,Y2,Color,Len,Orientation,Direction),
    equal(X_dir,0),
    greaterthan(Y_dir,0),
    translate(point(X,Y,C),point(X11,Y11,C11),X_dir1,Y_dir1,Color_obj2),
    linear_combination(Y_dir,Y_dir1,2,0),
    translate(line(X1,Y1,X2,Y2,Color,Len,Orientation,Direction),Line,X_dir,Y_dir).

target(line(X1,Y1,X2,Y2,Color,Len,Orientation,Direction),Line,X_dir,Y_dir):-
    point(X,Y,C),
    point(X11,Y11,C11),
    line(X1,Y1,X2,Y2,Color,Len,Orientation,Direction),
```

```
    equal(X_dir,0),
    greaterthan(Y_dir,0),
    translate(point(X,Y,C),point(X11,Y11,C11),X_dir1,Y_dir1,Color_obj2),
    linear_combination(Y_dir,Y_dir1,2,0),
    translate(line(X1,Y1,X2,Y2,Color,Len,Orientation,Direction),Line,X_dir,Y_dir).

target(line(X1,Y1,X2,Y2,Color,Len,Orientation,Direction),Line,X_dir,Y_dir):-
    point(X,Y,C),
    point(X11,Y11,C11),
    line(X1,Y1,X2,Y2,Color,Len,Orientation,Direction),
    equal(X_dir,0),
    greaterthan(Y_dir,0),
    translate(point(X,Y,C),point(X11,Y11,C11),X_dir1,Y_dir1,Color_obj2),
    linear_combination(Y_dir,Y_dir1,2,0),
    translate(line(X1,Y1,X2,Y2,Color,Len,Orientation,Direction),Line,X_dir,Y_dir).

target(line(X1,Y1,X2,Y2,Color,Len,Orientation,Direction),Line,X_dir,Y_dir):-
    point(X,Y,C),
    point(X11,Y11,C11),
    line(X1,Y1,X2,Y2,Color,Len,Orientation,Direction),
    equal(X_dir,0),
    greaterthan(Y_dir,0),
    translate(point(X,Y,C),point(X11,Y11,C11),X_dir1,Y_dir1,Color_obj2),
    linear_combination(Y_dir,Y_dir1,2,0),
    translate(line(X1,Y1,X2,Y2,Color,Len,Orientation,Direction),Line,X_dir,Y_dir).
```

A.5

Task 150deff5

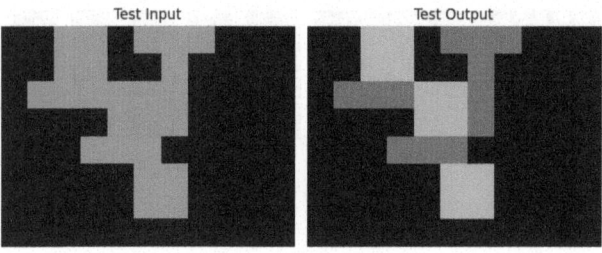

```
target(rectangle(X1,Y1,X2,Y2,X3,Y3,X4,Y4,Color,Clean,Area),Rectangle,Color_out,Cleanout):-
    rectangle(X1,Y1,X2,Y2,X3,Y3,X4,Y4,Color,Clean,Area),
    equal(Area,4),
    equal(Color_out,'blue'),
    equal(Cleanout,Clean),
    copy(rectangle(X1,Y1,X2,Y2,X3,Y3,X4,Y4,Color,Clean,Area),Rectangle,Color_out,Cleanout).

target(line(X1,Y1,X2,Y2,Color,Len,Orientation,Direction),Line,Color_out,Cleanout):-
    line(X1,Y1,X2,Y2,Color,Len,Orientation,Direction)
    equiv(Len,3),
    equal(Color_out,'red'),
    copy(line(X1,Y1,X2,Y2,Color,Len,Orientation,Direction),Line,Color_out,Cleanout).
```

References

1. Chollet, F.: On the measure of intelligence. arXiv preprint arXiv:1911.01547 (2019)
2. Lee, S., et al.: Reasoning abilities of large language models: in-depth analysis on the abstraction and reasoning corpus. arXiv preprint arXiv:2403.11793 (2024)
3. Butt, N., et al.: CodeIt: Self-Improving Language Models with Prioritized Hindsight Replay. arXiv preprint: arXiv:2402.04858 (2024)
4. Mikel, B.-I., Banerjee, S.: Neural networks for abstraction and reasoning: towards broad generalization in machines. arXiv preprint: arXiv:2402.03507 (2024)
5. Xu, Y., Li, W., Vaezipoor, P., Sanner, S., Khalil, E.B.: LLMS and the abstraction and reasoning corpus: Successes, failures, and the importance of object-based representations. arXiv preprint arXiv:2305.18354 (2023)
6. Lei, C., Lipovetzky, N., Ehinger, K.A.: Generalized Planning for the Abstraction and Reasoning Corpus. arXiv preprint arXiv:2401.07426 (2024)
7. Mitchell, M., Palmarini, A.B., Moskvichev, A.: Comparing Humans, GPT-4, and GPT-4V on abstraction and reasoning tasks. arXiv preprint: arXiv:2311.09247 (2023)
8. Johnson, A., Vong, W.K., Lake, B.M., Gureckis, T.M.: Fast and flexible: Human program induction in abstract reasoning tasks. arXiv preprint: arXiv:2103.05823 (2021)
9. Acquaviva, S., et al.: Communicating natural programs to humans and machines. In: Advances in Neural Information Processing Systems, vol. 35, pp. 3731–3743 (2022)
10. Kirchheim, K., Gonschorek, T., Ortmeier, F.: Out-of-Distribution Detection With Logical Reasoning. In: Proceedings of the IEEE/CVF Winter Conference on Applications of Computer Vision, pp. 2122–2131 (2024)
11. Farquhar, S., Gal, Y.: What 'Out-of-distribution. Is and Is Not. In: NeurIPS ML Safety Workshop (2022)
12. Ye, N., et al.: OoD-bench: quantifying and understanding two dimensions of out-of-distribution generalization. In: Proceedings of the IEEE/CVF Conference on Computer Vision and Pattern Recognition, pp. 7947–7958 (2022)
13. Spelke, E.S., Kinzler, K.D.: Core knowledge. Dev. Sci. **10**(1), 89–96 (2007)
14. Witt, J., Rasing, S., Dumančić, S., Guns, T., Carbon, C.-C.: A Divide-Align-Conquer Strategy for Program Synthesis. arXiv preprint: arXiv:2301.03094 (2023)
15. Raven, J.: Raven progressive matrices. In: McCallum, R.S. (eds.) Handbook of Nonverbal Assessment, pp. 223–237. Springer, Boston, MA (2003). https://doi.org/10.1007/978-1-4615-0153-4_11
16. Marcus, G.: Deep learning: a critical appraisal. arXiv preprint arXiv:1801.00631 (2018)
17. Singh, M., Cambronero, J., Gulwani, S., Le, V., Verbruggen, G.: Assessing GPT4-V on Structured Reasoning Tasks. arXiv preprint: arXiv:2303.08774 (2023)
18. Achiam, J., et al.: Gpt-4 technical report. arXiv preprint arXiv:2303.08774 (2023)
19. Lake, B.M., Ullman, T.D., Tenenbaum, J.B., Gershman, S.J.: Building machines that learn and think like people. Behav. Brain Sci. **40**, e253 (2017)
20. LeCun, Y., Bengio, Y., Hinton, G.: Deep learning. Nature **521**(7553), 436–444 (2015)
21. Cropper, A., Dumančić, S., Evans, R., Muggleton, S.H.: Inductive logic programming at 30. Mach. Learn. **111**(1), 147–172 (2022)
22. Muggleton, S., De Raedt, L.: Inductive logic programming: theory and methods. J. Logic Program. **19**, 629–679 (1994)

23. Gulwani, S., Hernández-Orallo, J., Kitzelmann, E., Muggleton, S.H., Schmid, U., Zorn, B.: Inductive programming meets the real world. Commun. ACM **58**(11), 90–99 (2015)
24. Gulwani, S., Polozov, O., Singh, R., et al.: Program synthesis. Found. Trends® Program. Lang. **4**(1–2), 1–119 (2017)
25. Carbonell, J.G., Michalski, R.S., Mitchell, T.M.: An overview of machine learning. Mach. Learn. **1**, 3–23 (1983)
26. Lin, D., Dechter, E., Ellis, K., Tenenbaum, J., Muggleton, S.: Bias reformulation for one-shot function induction. Front. Artif. Intell. App. **263**, 525–530 (2014)
27. Muggleton, S., Dai, W.-Z., Sammut, C., Tamaddoni-Nezhad, A., Wen, J., Zhou, Z.-H.: Meta-interpretive learning from noisy images. Mach. Learn. **107**, 1097–1118 (2018)
28. Körner, P., et al.: Fifty years of Prolog and beyond. Theory Pract. Logic Program. **22**(6), 776–858 (2022)
29. Cropper, A.: Learning logic programs through divide, constrain, and conquer. Proc. AAAI Conf. Artif. Intell. **36**, 6446–6453 (2022)
30. Quinlan, J.R.: Learning logical definitions from relations. Mach. Learn. **5**, 239–266 (1990)
31. Naveed, H., et al.: A comprehensive overview of large language models. arXiv preprint arXiv:2307.06435 (2023)
32. Guo, Z., et al.: Evaluating large language models: a comprehensive survey. arXiv preprint arXiv:2310.19736 (2023)
33. Shrestha, A., Mahmood, A.: Review of deep learning algorithms and architectures. IEEE Access **7**, 53040–53065 (2019)
34. Wu, Z., et al.: Reasoning or reciting? exploring the capabilities, limitations, and societal implications of large language models. arXiv preprint arXiv:2304.01373 (2023)
35. Rae, J.W., et al.: Scaling language models: Methods, analysis and insights from training Gopher. arXiv preprint: arXiv:2112.11446 (2021)
36. McCoy, R.T., Yao, S., Friedman, D., Hardy, M., Griffiths, T.L.: Embers of autoregression: understanding large language models through the problem they are trained to solve. arXiv preprint: arXiv:2309.13638 (2023)

Trustworthy Inductive Knowledge for Tropical Cyclones Formation Detection

Marco Adriano Ferrero[1]($\boxtimes$)[📷], Davide Azzalini[1][📷], Matteo Giuliani[1][📷], Andrea Castelletti[1][📷], Federico Cerutti[2][📷], and Francesco Amigoni[1][📷]

[1] Politecnico di Milano, piazza Leonardo da Vinci 32, 20133 Milano, Italy
marcoadriano.ferrero@mail.polimi.it,
{davide.azzalini,matteo.giuliani,andrea.castelletti,
francesco.amigoni}@polimi.it
[2] Università degli Studi di Brescia, piazza del Mercato 15, 25121 Brescia, Italy
federico.cerutti@unibs.it

Abstract. Predicting extreme weather events requires a careful mix of inductive and deductive reasoning as well as an understanding of epistemic and aleatory uncertainties to inform effective decision-making in high-risk situations. Tropical cyclones are among the most devastating extreme events and can cause extensive environmental damage, especially when they reach coastlines and inhabited areas due to their wide geographic extent and destructive power. This paper provides insight into two white-box Machine Learning methods applied to tropical cyclone forecasting, focusing on the interpretability of their predictions and their ability to handle uncertainty: both aspects are fundamental for supporting any successive deductive inference and decision-making activity. We consider Decision Trees and Bayesian Rule Lists, which demonstrate effectiveness in describing the presence of a cyclone and achieving mutually consistent results, and we critically assess their strengths and weaknesses.

Keywords: Tropical Cyclones · Explainable Artificial Intelligence · Decision Trees · Bayesian Rule Lists

1 Introduction

Decision-making in high-risk scenarios, such as predicting extreme weather events, necessitates a meticulous blend of inductive and deductive reasoning. Inductive reasoning helps form models from specific data, while deductive reasoning applies general laws to predict outcomes. Moreover, recognising epistemic uncertainty, or the unknowns in our knowledge, allows for refining these models. Meanwhile, aleatory uncertainty, the inherent randomness in weather, requires prediction flexibility [4,7,8].

Tropical Cyclones (TCs) are some of Earth's most extreme weather phenomena, characterised by strong winds, torrential rains, and storm surges [23].

They are defined as intense low-pressure systems that gain energy from warm water near the equator and rotate over tropical oceans. Their impact on coastlines and population centres can be devastating, causing valuable environmental and economic damages [3].

TC forecasting aims to detect the occurrence and development of cyclonic activity in particular areas of our planet. From the last century until today, climatologists and scientists have devised various solutions to this problem [20]. However, due to these phenomena' extreme complexity and irregularity, it is still difficult to fully answer some open questions regarding the genesis, intensity, track, and extreme events associated with TCs [5]. In this context, Machine Learning (ML) techniques, which aim to learn relationships and patterns within data, have proven to be highly impactful in contributing to more accurate forecasting systems [21] (see Appendix 2 for a review). However, ML-based solutions are often difficult to interpret.

In this paper, we investigate the use of interpretable ML models to forecast the genesis of tropical cyclones. In particular, we analyse the inductive process of learning models of TC formation from meteorological data (Appendix 4) through the lenses of the knowledge that a human decision-maker can harness from it. We focus (Appendix 3) on Decision Trees [1] and Bayesian Rule Lists [13], and show that they demonstrate effectiveness in describing the presence of a cyclone, achieve mutually consistent results, and – in particular for Bayesian Rule Lists – provide a sophisticate evaluation of the associated epistemic and aleatory uncertainty (Appendix 5). We argue that integrating inductive and deductive approaches can lead to a better understanding of the relationships between meteorological data and TCs' activities. In this sense, we promote accurate and transparent forecasting systems for TCs that support human analysts in understanding the phenomena and making the decision-making process more transparent.

2 Related Work

Tropical cyclones are dramatically complex atmospheric anomalies with intricate dynamic processes influenced by several meteorological and oceanic elements. Over time, employing model-based (*e.g.*,equation-based) approaches to represent TCs has shown various limitations, mainly linked to the difficulties of modelling such large systems [22]. Therefore, using data-driven approaches has opened up new possibilities, showing general improvements over the four main problems related to cyclone forecast: genesis, intensity, tracking, and related weather destructive events [24]. Since in this paper we are concerned with forecasting the genesis of cyclones, we focus the following discussion on this topic.

Tropical cyclogenesis is defined as TC development from a pre-existing tropical disturbance. Various statistical and ML techniques, combined with more traditional model-based methods, are used for the prediction of these events but their accuracy is still an open problem.

Depending on the time window, long-term forecasting involves predicting cyclone formation over extended periods, while short-term forecasting deals with predicting cyclone emergence within a short period. We focus on the short-term problem.

In this context, Zhang et al. [29] propose using decision trees [1] (see also Appendix 3). The result is a straightforward model with a tree depth of five. Its structure facilitates interpretation by clearly defining the variables and their thresholds relevant to predicting the phenomenon. Influential variables include relative vorticity at 800 hPa, Sea Surface Temperature (SST), precipitation rate, mean divergence between 1000 and 500 hPa, and air temperature at 300 hPa.

Zhang et al. [30] provide further insight, analysing how different models behave in predicting the development of a Mesoscale Convective System (MCS) – a collection of thunderstorms that act as a system – into a tropical cyclone, evaluating performance at various times and assessing the importance of the variables used. The dataset employed for this work includes 13 environmental variables useful for describing MCSs and TCs.

Building on previous research about interpretable[1] models for predicting TCs, we aim to assess the inductively derived knowledge and uncertainty predictions in light of using such knowledge for deductive decision-making. To achieve this, we employ two white-box ML techniques: Decision Trees and Bayesian Rule Lists (Appendix 3). These methods were chosen for their inherent interpretability and ability to handle – albeit very differently – different forms of uncertainty, allowing us to gain clear, actionable insights into TCs forecasting within the specified geographical area.

We are clearly aware of further developments using deep learning methods, *e.g.*,[16], which, however, lacks the trait of interpretability we seek to investigate in this paper.

3 Methodology

This paper uses supervised learning techniques to approach the TCs forecasting as a binary classification problem. The learned models take as input the 16 meteorological variables and the temporal indicator (Appendix 4) and produce a binary output that indicates the presence or absence of a cyclone. The data input consists of the values of the 17 attributes relative to the previous day. Forecasts have varying time horizons, ranging from one to ten days after the date corresponding to the input sample.

Concerning the models used for supervised learning, we focus on Decision Trees [1] and Bayesian Rule List (BRL) [13]. BRL generates a posterior distribution over permutations of if-then rules from a vast set of potential rules, yielding interpretable decision lists that allow for an estimation of epistemic

---

[1] Interpretability refers to the ability to understand a ML model and explain its decision-making process in a way that humans can comprehend, see [14] for a more in-depth discussion.

and aleatory uncertainty. This is due to a hierarchical prior that prioritises succinctness in the number of rules and the complexity of individual rules. BRL balances accuracy, interpretability, and computational demand, addressing the challenge of creating predictive models with discrete input space akin to decision trees, lists, or associative classifiers. It offers a solution that avoids the pitfalls of greedy methods like CART or C5.0, which can compromise solution quality, by finding a middle ground in optimising Decision Trees.

3.1 Decision Trees

Algorithms building Decision Trees [1] break greedily down a dataset into smaller subsets while incrementally creating classification rules. The final result is a tree with decision nodes and leaf nodes: a decision node has two or more branches, each representing values for the attribute tested, and a leaf node represents a decision taken after computing all attributes. The paths from root to leaves represent classification rules. Learning a decision tree involves selecting attributes that best partition the data, typically using criteria like information gain.

Estimating uncertainty in decision trees often relies on methods like bootstrapping, cross-validation, or ensemble techniques, which provide empirical approximations. However, these approaches lack the probabilistic foundation of Bayesian methods, which naturally quantify uncertainty through prior distributions and observed data. This can lead to more robust and interpretable uncertainty estimates in Bayesian models compared to the potentially less reliable and more biased estimates derived from decision tree techniques.

3.2 Bayesian Rule Lists

The general setting for BRL [13] is multi-class classification, where the set of possible labels is $1, \ldots, L$. The training data are pairs $\{(x_i, y_i)\}_{i=1}^n$, where $x_i \in \mathbb{R}^d$ are the features of observation i, and y_i are the labels, $y_i \in \{1, \ldots, L\}$. We let $\boldsymbol{x} = (x_1, \ldots, x_n)$ and $\boldsymbol{y} = (y_1, \ldots, y_n)$.

An association rule $a \rightarrow b$ is an implication with an antecedent a and a consequent b. For classification, the antecedent is an assertion about the feature vector x_i that is either true or false, for example, "$x_{i,1} = 1$ and $x_{i,2} = 0$." This antecedent contains two conditions, which we call the cardinality of the antecedent. The consequent b would typically be a predicted label y. A Bayesian association rule has a multinomial distribution over labels as its consequent rather than a single label:

$$a \rightarrow y \sim \text{Multinomial}(\boldsymbol{\theta}). \tag{1}$$

The multinomial probability is then given a prior, leading to a *prior consequent distribution*:

$$\boldsymbol{\theta}|\boldsymbol{\alpha} \sim \text{Dirichlet}(\boldsymbol{\alpha}). \tag{2}$$

Table 1. An example of a Bayesian decision list.

if	a_1	$y \sim$ Multinomial$(\boldsymbol{\theta}_1)$, $\boldsymbol{\theta}_1 \sim$ Dirichlet$(\boldsymbol{\alpha} + \boldsymbol{N}_1)$
else if	a_2	$y \sim$ Multinomial$(\boldsymbol{\theta}_2)$, $\boldsymbol{\theta}_2 \sim$ Dirichlet$(\boldsymbol{\alpha} + \boldsymbol{N}_2)$
$\vdots$		
else if	a_m	$y \sim$ Multinomial$(\boldsymbol{\theta}_m)$, $\boldsymbol{\theta}_m \sim$ Dirichlet$(\boldsymbol{\alpha} + \boldsymbol{N}_m)$
else		$y \sim$ Multinomial$(\boldsymbol{\theta}_0)$, $\boldsymbol{\theta}_0 \sim$ Dirichlet$(\boldsymbol{\alpha} + \boldsymbol{N}_0)$

Given observations $(\boldsymbol{x}, \boldsymbol{y})$ classified by this rule, we let $N_{.,l}$ be the number of observations with label $y_i = l$, and $\boldsymbol{N} = (N_{.,1}, \ldots, N_{.,L})$. We then obtain a *posterior consequent distribution*:

$$\boldsymbol{\theta} | \boldsymbol{x}, \boldsymbol{y}, \boldsymbol{\alpha} \sim \text{Dirichlet}(\boldsymbol{\alpha} + \boldsymbol{N}). \tag{3}$$

The core of a Bayesian decision list is an ordered antecedent list $d = (a_1, \ldots, a_m)$. Let $N_{j,l}$ be the number of observations x_i that satisfy a_j but not any of $a_1, \ldots, a_{j-1}$, and that have label $y_i = l$. This is the number of observations to be classified by antecedent a_j that have label l. Let $N_{0,l}$ be the number of observations that do not satisfy any of $a_1, \ldots, a_m$ and that have label l. Let $\boldsymbol{N}_j = (N_{j,1}, \ldots, N_{j,L})$ and $\boldsymbol{N} = (\boldsymbol{N}_0, \ldots, \boldsymbol{N}_m)$.

A Bayesian decision list $D = (d, \boldsymbol{\alpha}, \boldsymbol{N})$ is an ordered list of antecedents together with their posterior consequent distributions (see, for instance, Table 1). The posterior consequent distributions are obtained by excluding data that have satisfied an earlier antecedent in the list. Any observations that do not satisfy any of the antecedents in d are classified using the *default rule parameter* $\boldsymbol{\theta}_0$.

In building the Bayesian decision list, BRL selects antecedents from a pre-established collection. This is particularly applicable to data with binary or categorical attributes. The procedure involves using frequent itemset mining techniques, wherein the itemsets serve as the antecedents. In cases where features are binary, the *FP-growth* algorithm, as proposed by [2], is employed for mining antecedents. This algorithm efficiently discovers all itemsets that meet the specified constraints regarding minimum support and maximum cardinality.

Let $\mathcal{A}$ the set of antecedents, $a_{<j}$ the antecedents before j in the rule list if there are any, c_j be the cardinality of antecedent a_j, and $c_{<j}$ the cardinalities of the antecedents before j in the rule list. The generative model proposed in [13] samples from the posterior distribution of antecedent lists:

$$p(d|\boldsymbol{x}, \boldsymbol{y}, \mathcal{A}, \boldsymbol{\alpha}, \lambda, \eta) \propto p(\boldsymbol{y}|\boldsymbol{x}, d, \boldsymbol{\alpha}) p(d|\mathcal{A}, \lambda, \eta). \tag{4}$$

The hyperparameters $\boldsymbol{\alpha}$, λ, and η are user-defined.

The likelihood function follows directly from the generative model. Let $\boldsymbol{\theta} = (\boldsymbol{\theta}_0, \boldsymbol{\theta}_1, \ldots, \boldsymbol{\theta}_m)$ be the consequent parameters corresponding to each antecedent in d together with the default rule parameter $\boldsymbol{\theta}_0$.

Given the generative model, the likelihood function for parameters $\boldsymbol{\theta} = (\theta_0, \theta_1, \ldots, \theta_m)$ is:

$$p(\boldsymbol{y}|\boldsymbol{x}, d, \boldsymbol{\theta}) = \prod_{j: \sum_l N_{j,l} > 0} \text{Multinomial}(\mathbf{N}_j | \boldsymbol{\theta}_j), \tag{5}$$

with $\boldsymbol{\theta}_j$ following a Dirichlet distribution:

$$\boldsymbol{\theta}_j \sim \text{Dirichlet}(\boldsymbol{\alpha}). \tag{6}$$

The Dirichlet-multinomial distribution is obtained by:

$$p(\mathbf{y}|\mathbf{x}, d, \boldsymbol{\alpha}) \propto \prod_{j=0}^{m} \frac{\prod_{l=1}^{L} \Gamma(N_{j,l} + \alpha_l)}{\Gamma(\sum_{l=1}^{L} N_{j,l} + \alpha_l)}. \tag{7}$$

The uniform prior is set with $\boldsymbol{\alpha}$ where $\alpha_l = 1$ for all l.

4 Datasets

TCs are weather events that can form in various regions of the world, with slight variations and frequency depending on the basin in which they occur [12]. This study focuses on the South-West Indian Ocean basin, with geographical coordinates of 0°S, 30°S, 30°W, and 90°W, as a significant target area for forecasts.

The datasets considered for this work cover large-scale global drivers, local variables relative to the reference area, and data concerning the presence of TCs in the target area. All the data involved are historical observations collected daily from 1980 to 2022.

Global drivers concern seasonal or intraseasonal phenomena that occur regularly and are often related to the chances of observing TCs. For this study, we consider both the Madden-Julian Oscillation (MJO) and the El Niño-Southern Oscillation (ENSO). The MJO consists of large-scale coupled patterns in atmospheric circulation and deep convection that propagate slowly (5 m/s) Eastward across the warm sea surfaces of the Indian and Pacific Oceans [28]. A pair of Principal Components (PCs) time series comprise the index that describes this anomaly under the real-time multivariate MJO series, RMM1 and RMM2 [25]. ENSO instead refers to a quasi-periodic event that alters wind and SSTs in the Pacific Ocean, causing irregularities in tropical and subtropical areas [17]. The data used to describe this phenomenon represent SSTs in specific Pacific and Indian Ocean areas.

Local drivers are relative to the observations for meteorological variables occurring in the target region extracted from the ERA5 reanalysis dataset by the Copernicus Climate Change Service (C3S)[2]. This system retrieves data on local meteorological variables, such as daily weather conditions for wind speed, surface pressure, and precipitation. ERA5 provides data with 0.25° resolution in latitude or longitude. As a reanalysis system, it details the global atmosphere, land surface, and ocean waves from 1950 to the present [6].

---
[2] https://climate.copernicus.eu/climate-reanalysis (on 15th June 2024).

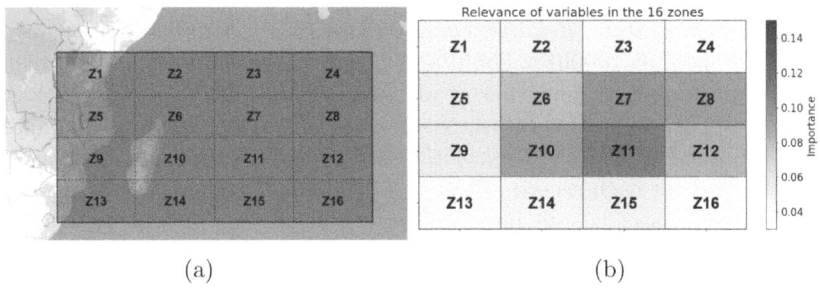

Fig. 1. Segmentation of the target region (1a) and zone importance (1b)

The third data type describes the *historical occurrences of TCs* in the area of interest. It is extracted from the International Best Track Archive for Climate Stewardship (IBTrACS) [11], a public dataset that provides information on the position, intensity, and size of TCs worldwide. The archive covers a period from the beginning of the previous century to the present day. These data are essential for labelling samples and implementing supervised methods.

The ERA5 reanalysis system offers high-resolution data for the target area. However, the high data dimensionality poses a challenge when training ML models [18,19]. Therefore, reducing the dimensionality of our data is necessary to minimise computing effort and eliminate irrelevant information to ease the interpretability of resulting models.

We segment the target region into a few sub-areas and evaluate the mean and standard deviation of the local drivers for each sub-area. To address the segmentation issue, Kernel Density Estimation (KDE) [26] is applied to determine the Probability Density Function (PDF) of relevant variables through data analysis. In particular, it evaluates how the meteorological variables distribute over the target region. PDFs are computed on different geographic segmentations (in 4, 9, 16, 25, and 36 sub-areas) of the basin to find the best compromise between an approximation of the original ERA5 data distribution and a reasonable reduction of its dimensionality. Comparing the different PDFs obtained, the optimal solution for this work results in segmenting the target region into 16 sub-areas, as displayed in Fig. 1a. Then, the Z11 sub-area is considered a new target zone. Z11 covers an ocean surface of more than one million km^2 with coordinates of 15°S, 22.5°S, 60°W, and 75°W. The choice of this particular zone is the result of the features selection (see also below) and the TCs distribution analysis (Fig. 1b). Additionally, the high number of TCs collected in Z11 by IBTrACS from 1980 to 2022 makes it a perfect candidate to have a rather balanced dataset.

Then, for the preliminary analysis reported in this paper, a further reduction of the local and global variables is carried out. Global drivers are not considered anymore, as their contribution to short-term forecasting problem is not very relevant. Moreover, only a subset of local variables is selected, according to a tree-based feature selection that computes features' importance based on the impurity evaluation. At the end, we consider 8 meteorological drivers (*Surface Pressure*,

Sea Surface Temperature, *Relative Vorticity* at 850hPa, *Air Density*, *Wind Gust Speed*, *Wind Speed* at 1000hPa, 850hPa, 300hPa). For each of them, we consider the mean and standard deviation of all data points in Z11 as meteorological contributors. Therefore, the dataset consists of 16 meteorological contributors and one temporal indicator (*Day of the Year*), which specifies the number of days in the year for each record.

The source dataset is split into two parts: the training set comprises all the samples covering 1980–2011, while the test set covers 2012–2022.

Table 2. Evaluation of the Decision Tree and Bayesian Rule Lists for 24-hour forecasts.

Metrics	DT	BRL
Accuracy	0.98	0.98
Precision	0.80	0.78
Recall	0.66	0.70
False Alarm rate	0.20	0.22

5 Results and Discussion

We implemented two solutions for TC forecasting, one using Decision Trees and the other using Bayesian Rule Lists. Here, we report solely the 24-hour forecast for TCs detection, as extending the prediction period is widely considered challenging [9]. The evaluation metrics for the two white-box models we implemented are presented in Table 2, with both models demonstrating comparable predictive performance within the metrics evaluated.

Table 3. Three most significant—based on the support—decision paths obtained by the Decision Tree model for the TCs forecasting. $\cdot_i$ is the i-th threshold used to discretise the feature considered in the inequality. The ratio is the number of positive examples divided by all examples covered by the decision path.

		Ratio	Support
if	$\mathbb{E}[Vor\_850hPa] \leq \cdot_1 \land \mathsf{SD}(Vor\_850hPa) > \cdot_2 \land \mathsf{SD}(Wind\_300hPa) \leq \cdot_1 \land \mathsf{SD}(P) > \cdot_2 \land \mathsf{SD}(T\_1000hPa) > \cdot_1$	0.96	231
if	$\mathbb{E}[Vor\_850hPa] > \cdot_1 \land \mathsf{SD}(Vor\_850hPa) > \cdot_4 \land \mathsf{SD}(T\_1000hPa) \leq \cdot_2 \land \mathbb{E}[Vor\_850hPa] \leq \cdot_1 \land \mathsf{SD}(Wind\_Gust) > \cdot_1$	0.88	52
if	$\mathbb{E}[Vor\_850hPa] > \cdot_1 \land \mathsf{SD}(Vor\_850hPa) \leq \cdot_4 \land \mathbb{E}[P] \leq \cdot_1 \land \mathsf{SD}(P) > \cdot_1 \land \mathsf{SD}(Vor\_850hPa) > \cdot_1$	0.61	44

Table 3 and Table 4 depict the top-three Decision Tree paths – ordered by decreasing support size – and the BRL obtained, respectively.

Let us stress that the ratio derived from a Decision Tree (see Table 3) does not represent a probability. Instead, it is simply a ratio of the number of positive examples to the total number of examples that would be classified by that particular rule. This ratio indicates the rule's ability to capture positive instances within the dataset; it, however, lacks a probabilistic foundation. Also, the epistemic uncertainty associated with the classification is only indirectly reflected by the support metric, which is the count of positive samples identified by the decision rule. The support measures the rule's applicability across the dataset, but it does not directly assess the uncertainty inherent in the classification decision.

In contrast, BRLs yield a more succinct set of decision rules than decision trees and provide a posterior Beta distribution. This distribution, the univariate equivalent of a Dirichlet distribution, offers a probabilistic interpretation of the rules. The Beta distribution quantifies the uncertainty of the classification by providing a range of probabilities, thus offering a more nuanced understanding of the model's predictions. The two parameters of a Beta distribution can be interpreted as the "weight of evidence" in favour (resp. against) of a certain proposition.

Figure 2 depicts the Beta distribution identified by the BRL – see Table 4 – that do not have a gargantuan weight of evidence. Indeed, each Beta distribution can be transformed into a linguistic expression focusing on its expected value and its variance. The expected value reflects the aleatory uncertainty, which is the inherent randomness or variability of the phenomenon under study. This type of uncertainty is intrinsic to the process and is typically irreducible. In contrast, the variance of the Beta distribution encapsulates the epistemic uncertainty. This uncertainty arises from incomplete knowledge or information about the system and can be reduced as more data become available, or the observer's understanding improves. While aleatory uncertainty is a fundamental property of the stochastic system, epistemic uncertainty reflects our current state of knowledge. Figure 2 thus focuses on the Beta distributions for which we don't have *total confidence* (see Table 4, last column).

Linguistic expressions have been generated adopting the proposal in [10, Section 3.7.2]. For this work, we relied on simple thresholding on the expected value and the variance of each beta distribution (Table 5) as follows. For the aleatory uncertainty, the labels are: *Absolutely not* ($\mathbb{E}[\beta.] \leq 0.001$), *Very unlikely* ($0.001 < \mathbb{E}[\beta.] < 0.15$), *Unlikely* ($0.15 \leq \mathbb{E}[\beta.] < 0.3$), *Somewhat unlikely* ($0.3 \leq \mathbb{E}[\beta.] < 0.45$), *Chances about even* ($0.45 \leq \mathbb{E}[\beta.] \leq 0.55$), *Somewhat likely* ($0.55 < \mathbb{E}[\beta.] \leq 0.7$), *Likely* ($0.7 < \mathbb{E}[\beta.] \leq 0.85$), *Very likely* ($0.85 < \mathbb{E}[\beta.] < 0.999$), *Absolutely* ($\mathbb{E}[\beta.] \geq 0.999$). For the epistemic uncertainty, the labels are: *No confidence* ($\text{Var}(\beta.) \geq 1 \times 10^{-1}$), *Low confidence* ($1 \times 10^{-2} < \text{Var}(\beta.) \leq 1 \times 10^{-1}$), *Some confidence* ($1 \times 10^{-3} < \text{Var}(\beta.) \leq 1 \times 10^{-2}$), *High confidence* ($1 \times 10^{-4} < \text{Var}(\beta.) \leq 1 \times 10^{-3}$), *Total confidence* ($\text{Var}(\beta.) \leq 1 \times 10^{-4}$).

Table 4. BRL for the TCs forecasting, where $\cdot_i$ represents the i-th threshold used to discretise the feature considered in the inequality.

if $\mathbb{E}[P] > \cdot_1 \wedge \text{SD}(Wind\_Gust) \leq \cdot_1$	$\beta_1 = \text{Beta}(15.32, 5090.26)$	Very unlikely with total confidence
else if $\mathbb{E}[P] > \cdot_1 \wedge \text{SD}(Vor\_850hPa) \leq \cdot_1$	$\beta_2 = \text{Beta}(6.78, 1123.60)$	Very unlikely with total confidence
else if $\text{SD}(P) \leq \cdot_1 \wedge \mathbb{E}[Vor\_850hPa] \geq \cdot_1$	$\beta_3 = \text{Beta}(32.30, 1436.08)$	Very unlikely with total confidence
else if $\text{SD}(Vor\_850hPa) \leq \cdot_1$	$\beta_4 = \text{Beta}(77.47, 646.57)$	Very unlikely with high confidence
else if $\mathbb{E}[P] > \cdot_1 \wedge \text{SD}(T\_1000hPa) > \cdot_2$	$\beta_5 = \text{Beta}(7.14, 111.81)$	Very unlikely with high confidence
else if $\mathbb{E}[Vor\_850hPa] \geq \cdot_1$	$\beta_6 = \text{Beta}(14.25, 39.74)$	Unlikely with some confidence
else if $\text{SD}(P) \leq \cdot_1$	$\beta_7 = \text{Beta}(41.94, 100.72)$	Unlikely with some confidence
else if $\mathbb{E}[Vor\_850hPa] \leq \cdot_1$	$\beta_8 = \text{Beta}(101.42, 80.34)$	Somewhat likely with some confidence
else if $\mathbb{E}[Wind\_850hPa] \leq \cdot_1$	$\beta_9 = \text{Beta}(101.61, 21.11)$	Likely with some confidence
else if $\text{SD}(P) > \cdot_2$	$\beta_{10} = \text{Beta}(145.39, 3.27)$	Very likely with high confidence
else	$\beta_{11} = \text{Beta}(1.32, 2.65)$	Somewhat unlikely with low confidence

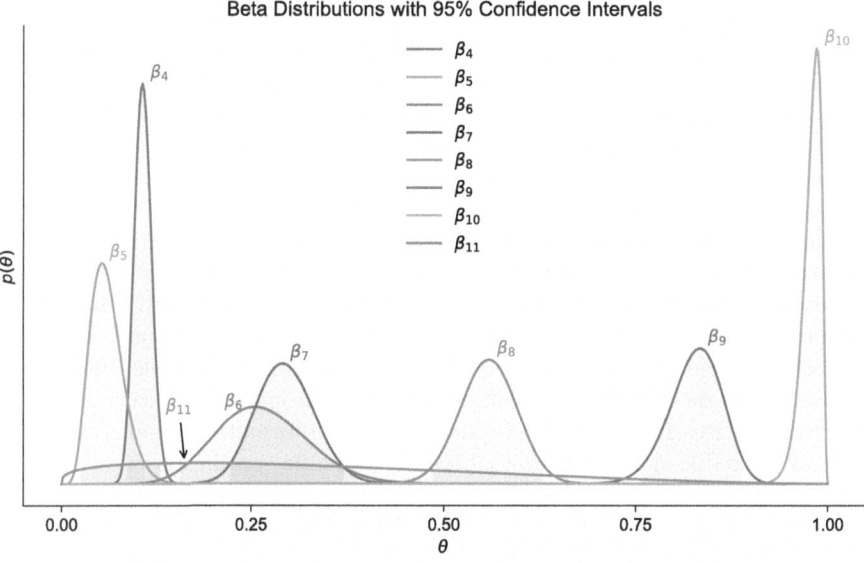

Fig. 2. Depiction of Beta distributions β_4, β_5, β_6, β_7, β_8, β_9, β_{10}, and β_{11} from the BRL for the TCs forecasting, see Table 4.

The simple thresholding employed in this paper is sufficient for demonstrating the basic concept. However, more sophisticated methods could be implemented to enhance the approach. These methods might include fuzzy labels [27], allowing for more nuanced and flexible classification.

The ability of BRL to generate such a nuanced understanding of the posterior epistemic and aleatory uncertainty improves our scientific understanding of a situation. Indeed, it is worth noticing that the scientific and the intelligence communities have long relied on scales for classifying pieces of information in terms of both likeabilities (*e.g., likely, very likely, ...*) and confidence (*e.g., low confidence, medium confidence, ...*), see for instance [15].

Table 5. Expected values and variances of the Beta distributions computed as per Table 4.

	Expected Value	Variance
β_1	0.003	5.86×10^{-7}
β_2	0.006	5.27×10^{-6}
β_3	0.022	1.46×10^{-5}
β_4	0.11	1.32×10^{-4}
β_5	0.06	4.70×10^{-4}
β_6	0.26	3.53×10^{-3}
β_7	0.29	1.44×10^{-3}
β_8	0.56	1.35×10^{-3}
β_9	0.83	1.15×10^{-3}
β_{10}	0.98	1.44×10^{-4}
β_{11}	0.33	4.47×10^{-2}

BRLs appear more interpretable and better suited for managing epistemic and aleatory uncertainty. However, this advantage comes at the cost of higher computational demand, taking 545 seconds compared to just 0.3 seconds for Decision Trees in our experiments. This increased complexity can be a limitation when computational resources or time are restricted.

Even though the experimental study has clear limitations, these do not undermine the paper's primary focus. While it lacks time series cross-validation techniques, analysis of learning and prediction times, hyperparameter tuning, and a broader range of machine learning algorithms, these shortcomings do not detract from the main objective. The study demonstrates how two white-box ML methods can improve tropical cyclone forecasting by enhancing interpretability and managing uncertainty. These aspects are crucial for transparent predictions and informed decision-making, which are the key contributions of this research.

6 Conclusion

Interpretability in ML-based systems, particularly those used for forecasting extreme weather events like TCs, is crucial for supporting decision-making. While data-driven approaches have shown increasing forecasting accuracy, many current systems lack transparency, hindering their utility. This paper contributes to assessing interpretable methods for TCs forecasting, providing descriptions of the predictions that meaningfully bind the input values to the TCs presence.

In the future, we will apply insights from these interpretable methods to improve decision-making in simulations with domain experts. This includes integrating the interpretability framework into exercises with meteorologists and emergency managers and conducting user studies to assess how model transparency affects decision accuracy and confidence in simulated extreme weather scenarios.

Acknowledgments. This work was partially supported by the European Office of Aerospace Research & Development (EOARD) under award number FA8655-22-1-7017 and by the Climate Intelligence (CLINT) project under H2020 framework programme of the European Union, grant number 101003876. Any opinions, findings, and conclusions or recommendations expressed in this material are those of the author(s) and do not necessarily reflect the views of the United States government.

References

1. Bishop, C.: Pattern recognition and machine learning. In: Springer-Verlag (2006)
2. Borgelt, C.: An implementation of the FP-growth algorithm. In: Proceedings Workshop on Open Source Data Mining: Frequent Pattern Mining Implementations, pp. 1–5 (2005)
3. Zehnder, J.: Tropical cyclone. In: Encyclopedia Britannica (2024). https://www.britannica.com/science/tropical-cyclone. Accessed 15 Jun 2024
4. Cerutti, F., Kaplan, L., Sensoy, M., et al.: Evidential reasoning and learning: a survey. In: Proceedings IJCAI, pp. 5418–5425 (2022)
5. Emanuel, K.: 100 years of progress in tropical cyclone research. In: Meteorological Monographs **59**, 15–1 (2018)
6. Hersbach, H., Bell, B., Berrisford, P., et al.: The ERA5 global reanalysis. In: Quarterly Journal of the Royal Meteorological Society, vol. 146, no. 730, pp. 1999–2049 (2020)
7. Hora, S.C.: Aleatory and epistemic uncertainty in probability elicitation with an example from hazardous waste management. In: Reliability Engineering & System Safety, vol. 54, no. 2, pp. 217–223 (1996)
8. Hüllermeier, E., Waegeman, W.: Aleatoric and epistemic uncertainty in machine learning: an introduction to concepts and methods. In: Machine Learning vol. 110, no. 3, pp. 457–506 (2021)
9. IFRC – International Federation of Red Cross And Red Crescent Societies. Mozambique, Cyclone Early Action Protocol Summary (EAP Number: EAP2023MZ03) (2024)
10. Josang, A.: Subjective logic: a formalism for reasoning under uncertainty. In: Springer (2016)
11. Knapp, K., Kruk, M., Levinson, D., et al.: The international best track archive for climate stewardship (IBTrACS) unifying tropical cyclone data. In: Bulletin of the American Meteorological Society, vol. 91, no. 3, pp. 363–376 (2010)
12. Lavender, S., McBride, J.: Global climatology of rainfall rates and lifetime accumulated rainfall in tropical cyclones: influence of cyclone basin, cyclone intensity and cyclone size. In: International Journal of Climatology, vol. 41, no. S1, pp. E1217–E1235 (2021)
13. Letham, B., Rudin, C., McCormick, T., et al.: Interpretable classifiers using rules and Bayesian analysis: building a better stroke prediction model. In: The Annals of Applied Statistics, vol. 9, no. 3, pp. 1350–1371 (2015)
14. Lipton, Z.: The mythos of model interpretability: in machine learning, the concept of interpretability is both important and slippery. In: Queue, vol. 16, no. 3, pp. 31–57 (2018)

15. Mastrandrea, M., Mach, K., Plattner, G.-K., et al.: The IPCC AR5 guidance note on consistent treatment of uncertainties: a common approach across the working groups. In: Climatic Change, vol. 108, no. 4, pp. 675 (2011)
16. Matsuoka, D., Nakano, M., Sugiyama, D., et al.: Deep learning approach for detecting tropical cyclones and their precursors in the simulation by a cloud-resolving global nonhydrostatic atmospheric model. In: Progress in Earth and Planetary Science, vol. 5, no. 1, pp. 1–16 (2018)
17. McPhaden, M., Zebiak, S., Glantz, M.: ENSO as an integrating concept in earth science. In: Science, vol. 314, no. 5806, pp. 1740–1745 (2006)
18. Molnar, C., Casalicchio, G., Bischl, B.: Interpretable machine learning: a brief history, state-of-the-art and challenges. In: Proceedings ECML PKDD, pp. 417–431 (2020)
19. Oates, T., Jensen, D.: The effects of training set size on decision tree complexity. In: Proceedings Workshop on Artificial Intelligence and Statistics, pp. 379–390 (1997)
20. Robertson, A.W., Vitart, F., Camargo, S.J.: Subseasonal to seasonal prediction of weather to climate with application to tropical cyclones. In: Journal of Geophysical Research: Atmospheres, vol. 125, no. 6, pp. e2018JD029375 (2020)
21. Salcedo-Sanz, S., Pérez-Aracil, J., Ascenso, G., et al.: Analysis, characterization, prediction, and attribution of extreme atmospheric events with machine learning and deep learning techniques: a review. In: Theoretical and Applied Climatology, vol. 155, no. 1, pp. 1–44 (2024)
22. Scoccimarro, E., Gualdi, S., Villarini, G., et al.: Intense precipitation events associated with landfalling tropical cyclones in response to a warmer climate and increased CO_2. In: Journal of Climate, vol. 27, no. 12, pp. 4642–4654 (2014)
23. Villarini, G., Goska, R., Smith, J.A., et al.: North Atlantic tropical cyclones and US flooding. In: Bulletin of the American Meteorological Society, vol. 95, no. 9, pp. 1381–1388 (2014)
24. Wang, Z., Zhao, J., Huang, H., et al.: A review on the application of machine learning methods in tropical cyclone forecasting. In: Frontiers in Earth Science, vol. 10, pp. 902596 (2022)
25. Wheeler, M., Hendon, H.: An all-season real-time multivariate MJO index: development of an index for monitoring and prediction. In: Monthly Weather Review, vol. 132, no. 8, pp. 1917–1932 (2004)
26. Węglarczyk, S.: Kernel density estimation and its application. In Proceedings of ITM Web of Conferences, vol. 23, pp. 00037 (2018)
27. Zadeh, L.A.: Fuzzy sets. In: Information and Control, vol. 8, no. 3, pp. 338–353 (1965)
28. Zhang, C.: Madden-Julian oscillation. In: Reviews of Geophysics, vol. 43, no. 2, pp. 2004RG000158 (2005)
29. Zhang, W., Fu, B., Peng, M., et al.: Discriminating developing versus nondeveloping tropical disturbances in the Western North Pacific through decision tree analysis. In: Weather and Forecasting, vol. 30, no. 2, pp. 446–454 (2015)
30. Zhang, T., Lin, W., Lin, Y., et al.: Prediction of tropical cyclone genesis from mesoscale convective systems using machine learning. In: Weather and Forecasting, vol. 34, no. 4, pp. 1035–1049 (2019)

Automatic Curriculum Cohesion Analysis Based on Knowledge Graphs

Paulina Gacek and Weronika T. Adrian(✉)

AGH University of Krakow, Al. A. Mickiewicza 30, 30-059 Krakow, Poland
paulinagacek@student.agh.edu.pl, wta@agh.edu.pl

Abstract. Creating a well-structured and cohesive curriculum is essential for providing students with a meaningful and organized educational experience. However, the complexity and scale of curricula and the need for regular updates make developing a curriculum for a single major a highly challenging task. In this paper, we aim to construct a comprehensive tool designed to automatically analyze existing curricula and identify areas of incoherence. The presented approach utilizes unsupervised learning techniques and large language models (LLMs) to extract the main concepts covered by courses and their prerequisites directly from course syllabi. The extracted entities are then linked to Wikidata entities, ensuring both their correctness and uniqueness. By constructing a knowledge graph based on the relationships between courses and main educational concepts, the cohesion of the curriculum can be systematically verified. The verification process involves checking if all cohesion logic predicates are satisfied, thereby ensuring that the curriculum is both comprehensive and logically structured. The results demonstrate the efficacy of our approach in identifying inconsistencies within educational curricula, thereby aiding in the development of more cohesive and well-organized educational programs.

Keywords: Knowledge graphs · Concept extraction · Curriculum design · Large language models

1 Introduction

Creating a well-structured and cohesive curriculum is paramount for providing students with a meaningful and organised educational experience. The design of the curriculum must consider the relationships between courses to ensure that students acquire the necessary knowledge and skills in a logical sequence. The inherent complexity of curricula, which often consist of numerous courses with interrelated content and prerequisites, poses significant challenges in ensuring coherence and cohesion. Common problems include redundancy (repeated coverage of topics across courses), gaps (prerequisite concepts not adequately addressed), and misalignment (essential concepts taught out of sequence). These issues can hinder the educational experience, leading to student confusion and fragmented knowledge.

In this work, we propose a novel tool for automatic systematic analysis of curriculum cohesion. The proposed method utilizes a two-stage approach. First, the tool extracts course content and prerequisites, linking them to a standardized knowledge base (e.g., Wikidata [14]). This extracted information is then used to construct a curriculum knowledge graph, enabling automatic verification of curriculum coherence. Knowledge graphs offer an intuitive and structured representation of entities and their relationships, allowing the creation of a network of knowledge structures based on the extracted data. By expressing cohesion violations (redundancy, gaps, and misalignment) as logical predicates and employing logical reasoning, the tool systematically checks the satisfiability of these predicates across the curriculum knowledge graph.

The paper is organized as follows: in Sect. 2, we outline the selected related research. In Sect. 3, we put forward the methodology of our approach, followed by an experimental evaluation in Sect. 4. The paper is concluded in Sect. 5.

2 Related Work

Several studies have explored the use of knowledge graphs in the context of curriculum design and educational systems, providing a foundation for understanding how knowledge graphs can enhance the organization and analysis of educational content.

The work by Yu et al. [11] proposes a construction scheme for a curriculum knowledge graph. The paper aimed to construct a knowledge graph to make further manual analysis easier. Their system includes nodes representing courses and main topics, with "prerequisite" being the only relationship type.

Chen et al. [12] introduce the KnowEdu system, which automatically constructs a knowledge graph capturing relationships between instructional concepts within a single course. They employ supervised machine learning with recurrent neural networks (RNNs) for concept extraction, trained on a pre-defined dataset. While effective within a course, KnowEdu's lack of modeling relationships between courses limits its applicability for curriculum analysis.

The study by Aldhalaan et al. [17] presents a system to construct a knowledge graph from a curriculum. Their system allows users to query the knowledge graph for information about courses and programs, aiding in the identification of redundancies and inconsistencies. However, similar to the previously mentioned works, this system focuses primarily on manual analysis and visualization, lacking automatic analysis functionalities.

Building upon these foundations, our work proposes a tool that automatically constructs a knowledge graph for a given curriculum. The approach further differentiates itself by performing an in-depth analysis of the curriculum's consistency and coherence, offering insights for educators and curriculum designers.

3 Methodology

In this section, we outline the methodology employed to construct our curriculum knowledge graph and utilize it for automatic cohesion analysis. The process is comprised of four key stages: ontology modeling, keyphrase extraction, entity filtering, and automatic cohesion analysis.

3.1 Ontology Modeling

A pre-defined ontology serves as the foundation for the knowledge graph, providing a structured vocabulary for representing relationships between courses and concepts across the curriculum. The knowledge graph consists of two classes of objects: `Course` and `Concept`, which are essential for performing cohesion checks. `Course` instances are described with the course name and the academic term in which the course is offered. `Concept` class represents an educational concept or academic discipline. Each concept is linked to a corresponding entity in the Wikidata knowledge base [14], using a unique identifier (ID). There are two types of relationships: `covers` and `prerequisite`, both of which connect courses to concepts. A course can cover multiple concepts and have multiple prerequisites. The resulting curriculum knowledge graph is stored in Neo4J [15], a graph database optimized for handling interconnected information.

3.2 Concepts Extraction

Course descriptions with program and prerequisites are typically found in their syllabi. Although the syllabi generally follow a predefined structure, including sections like Prerequisites, Learning outcomes, and Basic information, the lack of strict guidelines for these sections can lead to inconsistencies such as misspellings, missing key concepts, or the inclusion of irrelevant comments. Simple regular expressions or basic natural language processing tools are insufficient for accurately extracting the necessary information, hence the need for machine learning techniques. The proposed method leverages an unsupervised approach and, thus does not require any alterations to the architecture of pre-trained language models nor does it introduce any additional parameters, making it a simple yet potent approach for keyphrase extraction.

To differentiate between a course's prerequisites and its core concepts, keyphrases are extracted independently from both the prerequisites and program descriptions. This results in two distinct sets: concepts considered necessary prior knowledge for the course and key educational concepts covered by the course. To ensure comprehensive keyphrase extraction from lengthy course programme description, the document is divided into smaller sub-documents. The number of sub-documents is directly proportional to the course's ECTS (European Credit Transfer and Accumulation System) credit value. ECTS credits represent a standardized unit for measuring student workload across higher education institutions, reflecting the time and effort required to achieve a program's objectives.

Keyphrase Extraction Algorithm

The algorithm for extracting keyphrases from the provided document d can be divided into the following steps:

1. Candidate Keyphrases Generation

Given a single document d, a set of candidate keyphrases $C = c_1, c_2, ..., c_k$ is generated by splitting d based on a stop-word pattern. This is motivated by the observation that keyphrases often consist of multiple words and typically lack punctuation or stop words [6].

2. Document and Candidate Embeddings Creation

The algorithm creates embedding representations for both the document and each candidate keyphrase. A pre-trained BERT model [16] is employed for this purpose. These embeddings capture the semantic meaning of the text without requiring further fine-tuning of the model.

3. Candidates Probability Calculation

For each candidate keyphrase $c \in C$ a probability score, $prob$, is calculated to indicate the likelihood of it being a true keyphrase. This score is based on the number of words it contains, denoted by $word\ count(c)$. The function $prob(w)$ is a pre-defined function that captures the observed frequency distribution of phrase lengths within expert-annotated keyphrases from relevant datasets [1–3]. By leveraging this information, the function can assign higher probabilities to phrase lengths that are statistically more likely to represent keyphrases.

4. Scoring and Selection

A final score s is computed for each candidate $c \in C$. This score combines two factors: the cosine similarity between the candidate and document embeddings, and the previously calculated probability score. The cosine similarity metric reflects the semantic similarity between the candidate and the entire document. The final score is calculated using the following equation:

$$score(c) = cos\ similarity(embed(c), embed(d)) \times prob(word\ count(c)) \quad (1)$$

The top n candidates with the highest scores are selected as the keyphrases.

3.3 Entities Filtering and Linking

The keyphrase extraction algorithm used in the initial stages might generate terms that are not strictly educational concepts. To refine these entities, the Mistral Large model, a state-of-the-art Large Language Model (LLM), was employed. The model was tasked with filtering out terms that were overly general, unrelated to the course content, or action-oriented rather than representing educational concepts or academic disciplines.

After filtering, the identified educational concepts are queried against the Wikipedia knowledge base. For each term that matched a Wikipedia article, the corresponding entity in Wikidata is linked to the term, integrating the filtered concepts into a broader knowledge framework.

3.4 Automatic Cohesion Analysis

The application of the knowledge graph is crucial for detecting inconsistencies and violations in the curriculum. Several types of cohesion violations can occur:

1. Redundancy: A particular concept is covered by more than one course.
2. Gaps: A prerequisite concept is not covered by any course.
3. Misalignment: A prerequisite concept necessary for a course conducted during the nth term is covered by a course conducted during the mth term, where $m > n$.

To express these cohesion violations using predicate logic, the following predicates and variables are defined:

- $C(x, y)$: Concept x is covered by course y.
- $P(x, y)$: Concept x is a prerequisite for course y.
- $T(y, n)$: Course y is conducted during the nth term.

Using these predicates, the cohesion violations can be expressed as follows:

1. $\exists x\, \exists y_1\, \exists y_2\, (C(x, y_1) \wedge C(x, y_2) \wedge y_1 \neq y_2)$
2. $\exists x\, \exists y\, (P(x, y) \wedge \neg \exists z\, C(x, z))$
3. $\exists x\, \exists y\, \exists z\, \exists n\, \exists m\, (P(x, y) \wedge C(x, z) \wedge T(y, n) \wedge T(z, m) \wedge m > n)$

To detect the violations on the knowledge graph stored in the Neo4J database, the predicates were encoded in Cypher [13] query language, as shown below:

```
MATCH (x:Concept)<-[r:COVERS]-()
WITH x, count(r) as rs
WHERE rs > 1
RETURN x;

MATCH entity=(y:Course)-[p:PREREQUISITE]->(x:Concept)
WHERE NOT ()-[:COVERS]->(x)
RETURN entity;

MATCH w=(y1:Course)-[:PREREQUISITE]->(x:Concept),(y2:Course)
WHERE (y2)-[:COVERS]->(x) AND y2.term > y1.term
RETURN w, y2;
```

We acknowledge that some cohesion violations may be intentional, such as repeating the same topic across different courses. In such cases, the detected violation can be deliberately ignored by the syllabus authors. However, we believe it is beneficial to highlight these violations. If different aspects of the same topic are covered in these courses, this differentiation should be clearly reflected in the syllabus, to ensure that they will be classified as separate entities within the knowledge graph.

4 Experimental Evaluation

In this section, we present the experimental evaluation of our approach, focusing on the keyphrase extraction and concept extraction from the syllabi of AGH University of Krakow. We discuss the metrics and data, as well as demonstrate the cohesion detection analysis of the developed knowledge graph.

4.1 Keyphrase Extracion

To evaluate the proposed method for keyphrase extraction more comprehensively, three widely-adopted scientific publication datasets were used: Inspec [1], SemEval-2010 [2] and SemEval-2017 [3], with a combined total of 2,729 abstracts. The datasets were transformed to consist of abstracts and corresponding keyphrases that appear in the abstracts. Keyphrases whose stemmed forms were absent in the stemmed versions of the abstracts were removed to ensure the reliability of keyphrase extraction results comparison. Table 1 presents a summary of these datasets, including the number of words per abstract, the number of keyphrases per abstract, and the total number of abstracts.

Table 1. Statistics of the datasets used for evaluation.

	Words per abstract			Keyphrases per abstract			
	Min	Mean	Max	Min	Mean	Max	Abstracts
Inspec	15	125	502	1	9	33	1991
Sem-Eval-2017	60	170	355	2	12	30	500
Sem-Eval-2010	53	412	12479	1	5	20	238

The performance of the keyphrase extraction model was assessed by comparing F1 scores of the top N predicted keyphrases, denoted as $F1@N$. Given that our approach attempts to match a fixed number of outputs against a variable number of ground-truth keyphrases, the results may not accurately reflect the quality of extraction. Therefore, the additional metric $F1@\mathcal{O}$, introduced by Yuan et al. in [4], was employed. It compares ground-truth phrases with the top k predicted keyphrases, where k is the total number of ground-truth keyphrases. Before determining if two phrases match, the Porter stemmer [5] was applied to each word in the phrase. Subsequently, macro-averaging was utilized to aggregate the evaluation scores for all testing samples.

As baseline models, popular keyphrase extraction methods were selected: statistics-based methods such as RAKE [6] and YAKE [7], as well as graph-based methods such as TextRank [8] and PositionRank [9]. Table 2 presents the results of F1 scores for @10, @15, and @$\mathcal{O}$ of the proposed solution (KeyExtract) and the baseline models on the three benchmark datasets.

The results show that the proposed solution achieves the best performance on all evaluation metrics across all the datasets. We acknowledge that the scores

Table 2. Comparison of F1 scores for baseline models and the proposed model.

	Inspec			Sem-Eval-2010			Sem-Eval-2017		
	@10	@15	@$\mathcal{O}$	@10	@15	@$\mathcal{O}$	@10	@15	@$\mathcal{O}$
YAKE	0.17	0.2	0.19	0.2	0.15	0.12	0.16	0.19	0.17
RAKE	0.29	0.34	0.29	0.13	0.15	0.05	0.2	0.24	0.23
TextRank	0.26	0.3	0.3	0.21	0.24	0.24	0.23	0.27	0.29
PositionRank	0.23	0.26	0.27	0.18	0.24	0.24	0.2	0.24	0.27
KeyExtract	**0.38**	**0.43**	**0.38**	**0.33**	**0.33**	**0.31**	**0.28**	**0.31**	**0.35**

of the baseline models presented in other publications might not be completely comparable with the approach presented in this work due to the use of additional post-processing and filtering methods.

4.2 Main Concepts Extraction

The data used for main concept extraction originated from the curriculum of 'Computer Science and Intelligent Systems' offered at AGH University of Krakow. To leverage a pre-trained English language model, the important sections, namely Basic information, Program details, Prerequisites, and Learning outcomes, were extracted and translated into English.

Figures 1 and 2 illustrate the main concepts extracted from a fragment of the 'Computer Architectures' and 'Logic for Software Engineers' program descriptions. The figures display program fragments with keyphrases produced by the keyphrase extraction algorithm underlined. Educational concepts identified by the Large Language Model (LLM) are additionally bolded, and concepts that have corresponding entities in Wikidata, highlighted in the figure, are used as the building blocks for constructing the curriculum knowledge graph. It can be observed that each step of the main concept extraction process progressively filters out the less relevant information. This highlights the importance of each step in ensuring the final selection of high-quality entities for the knowledge graph.

Analysis and design of <u>combinational</u> and **<u>sequential logic</u>**, starting from the **<u>NAND gate</u>** and the **<u>JK flip-flop</u>** and reaching the design of an example computer with hardware and <u>microprogrammed control</u> unit. <u>Designing circuits</u> and computers using **<u>VHDL</u>**. Analysis and design of **<u>combinational logic</u>**: <u>logic gates</u> (AND, OR, NOT, NAND, NOR), **<u>arithmetic logic</u>**, encoders, decoders, multiplexers, demultiplexers, ROM memories.

Fig. 1. Main concept extraction for the fragment of the 'Computer architecture' program.

Basics of **logic**: **formal languages**, **syntax**, **semantics**, **inference**, **systems modeling** - and many others. Discussion and explanation of the subject/ course syllabus during the first class. Discussion of teaching content and supporting materials (literature, websites). Lecture content: Logic, logic and logic again. In particular: Introduction to logic, the essence of logic, the role and tasks of logic, areas of application, limitations. The role and importance of language. Language and formal language. Syntax, semantics, interpretation, model. **Logical properties**.

Fig. 2. Main concept extraction for the fragment of the 'Logic for software engineers' program.

4.3 Cohesion Violation Detection

The curriculum knowledge graph, constructed from the extracted main concepts and course information, facilitates the automatic detection of various cohesion violations. Figures 3, 4, and 5 illustrate examples of such violations identified by the proposed tool.

Figure 3 exemplifies a Type 1 violation (redundancy). The concept of 'Flow Network' appears as a learning outcome in both the 'Algorithms and Data Structures' and 'Graph Theory' courses. This redundancy can lead to inefficient use of student time and potential confusion if the content is not well-differentiated between courses.

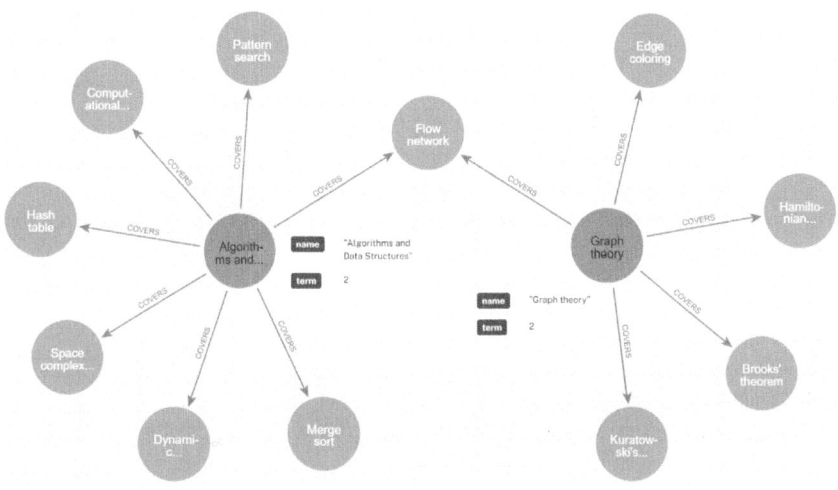

Fig. 3. Example of a Type 1 cohesion violation.

Figure 4 illustrates a Type 2 violation (gap). The concept of 'Python' is a prerequisite for both the 'Algorithms and Data Structures' and 'Operations Research' courses, but it is not explicitly covered as a learning outcome in any

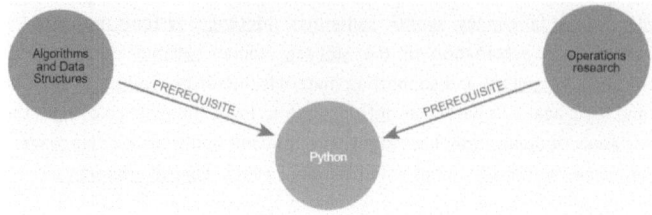

Fig. 4. Example of a Type 2 cohesion violation.

course. This gap can hinder student understanding in subsequent courses that rely on Python knowledge.

Finally, Fig. 5 depicts a Type 3 violation (misalignment). The concepts 'web API', 'front end', and 'client-server' are listed as prerequisites for the 'Software Engineering' course (offered in term 3). However, the knowledge graph reveals that these concepts are only covered in the 'Web Application Programming' course offered in term 5. This misalignment can create confusion for students who lack the necessary prior knowledge when studying Software Engineering.

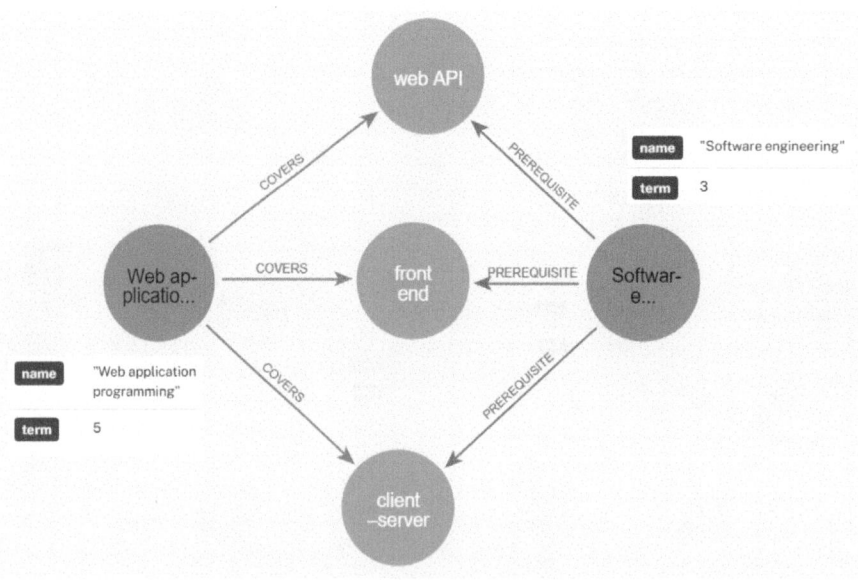

Fig. 5. Example of a Type 3 cohesion violation.

5 Conclusion

In this work, we proposed a tool for the automatic analysis of curriculum cohesion, leveraging the synergy between machine learning and symbolic graph-based knowledge representation. Our results demonstrate the efficacy of this approach in identifying and addressing inconsistencies within educational curricula, thereby contributing to the development of more cohesive and well-organized educational programs.

In the future, we plan to focus on detecting additional types of violations, such as identifying *similar* concepts covered by different courses and incorporating more complex relationships among the courses and concepts, such as hierarchical prerequisites. Additionally, connection to the Wikidata could be further leveraged by (i) exploring related concepts, and (ii) employing logical reasoning beyond queries, such that implicit knowledge could be used to ensure an even more comprehensive educational experience for students.

References

1. Hulth, A.: Improved automatic keyword extraction given more linguistic knowledge. In: Proceedings of the 2003 Conference on Empirical Methods in Natural Language Processing, pp. 216–223 (2003). https://aclanthology.org/W03-1028
2. Kim, S.N., Medelyan, O., Kan, M.-Y., Baldwin, T.: SemEval-2010 task 5: automatic keyphrase extraction from scientific articles. In: Proceedings of the 5th International Workshop on Semantic Evaluation, pp. 21–26, Uppsala, Sweden (2010). Association for Computational Linguistics. https://aclanthology.org/S10-1004
3. Augenstein, I., Das, M., Riedel, S., Vikraman, L., McCallum, A.: SemEval 2017 Task 10: ScienceIE - extracting keyphrases and relations from scientific publications. In: Proceedings of the 11th International Workshop on Semantic Evaluation (SemEval-2017), pp. 546–555, Vancouver, Canada (2017). Association for Computational Linguistics. https://doi.org/10.18653/v1/S17-2091, https://aclanthology.org/S17-2091
4. Yuan, X., et al.: One size does not fit all: generating and evaluating variable number of keyphrases. In: Proceedings of the 58th Annual Meeting of the Association for Computational Linguistics, pp. 7961–7975, Online (2020). Association for Computational Linguistics. https://doi.org/10.18653/v1/2020.acl-main.710, https://aclanthology.org/2020.acl-main.710
5. Porter, M.F.: An Algorithm for Suffix Stripping. In: Program, vol. 14, pp. 130–137 (1980). https://doi.org/10.1108/eb046814
6. Rose, S., Engel, D., Cramer, N., Cowley, W.: Automatic keyword extraction from individual documents. In: Text Mining: Applications and Theory, pp. 1–20 (2010). https://doi.org/10.1002/9780470689646.ch1
7. Campos, R., Mangaravite, V., Pasquali, A., Jorge, A., Nunes, C., Jatowt, A.: YAKE! Keyword extraction from single documents using multiple local features. In: Information Sciences **509**, 257–289 (2020). https://doi.org/10.1016/j.ins.2019.09.013, https://www.sciencedirect.com/science/article/pii/S0020025519308588
8. Mihalcea, R., Tarau, P.: TextRank: bringing order into text. In: Proceedings of the 2004 Conference on Empirical Methods in Natural Language Processing, pp. 404–411, Barcelona, Spain (2004). Association for Computational Linguistics, https://aclanthology.org/W04-3252

9. Florescu, C., Caragea, C.: PositionRank: an unsupervised approach to keyphrase extraction from scholarly documents. In: Proceedings of the 55th Annual Meeting of the Association for Computational Linguistics (Volume 1: Long Papers), pp. 1105–1115, Vancouver, Canada (2017). Association for Computational Linguistics. https://doi.org/10.18653/v1/P17-1102, https://aclanthology.org/P17-1102
10. Mendes, P.N., Jakob, M., García-Silva, A., Bizer, C.: DBpedia spotlight: shedding light on the web of documents. In: Proceedings of the 7th International Conference on Semantic Systems, pp. 1–8, Graz, Austria (2011). Association for Computing Machinery. https://doi.org/10.1145/2063518.2063519.
11. Yu, X., Stahr, M., Chen, H., Yan, R.G.: Design and implementation of curriculum system based on knowledge graph. In: 2021 IEEE International Conference on Consumer Electronics and Computer Engineering (ICCECE), pp. 767–770 (2020). https://api.semanticscholar.org/CorpusID:229363688
12. Chen, P., Yu, L., Zheng, V.W., Chen, X., Yang, B.: KnowEdu: a system to construct knowledge graph for education. In: IEEE Access vol. 6, pp. 31553–31563 (2018). https://doi.org/10.1109/ACCESS.2018.2839607
13. Francis, N., et al.: Cypher: an evolving query language for property graphs. In: Proceedings of the 2018 International Conference on Management of Data, pp. 1433–1445, Houston, TX, USA (2018). Association for Computing Machinery. https://doi.org/10.1145/3183713.3190657.
14. Vrandečić, D., Krötzsch, M.: Wikidata: a free collaborative knowledgebase. In: Communications of the ACM, vol. 57, no. 10, pp. 78–85 (2014). https://doi.org/10.1145/2629489
15. Webber, J.: A Programmatic Introduction to Neo4j. In: Proceedings of the 3rd Annual Conference on Systems, Programming, and Applications: Software for Humanity, pp. 217–218, Tucson, Arizona, USA (2012). Association for Computing Machinery. https://doi.org/10.1145/2384716.2384777.
16. Devlin, J., Chang, M.-W., Lee, K., Toutanova, K.: BERT: pre-training of deep bidirectional transformers for language understanding. In: Proceedings of the 2019 Conference of the North American Chapter of the Association for Computational Linguistics: Human Language Technologies, Volume 1 (Long and Short Papers), pp. 4171–4186, Minneapolis, Minnesota (2019). Association for Computational Linguistics. https://doi.org/10.18653/v1/N19-1423, https://aclanthology.org/N19-1423
17. Aldhalaan, G., Berri, J.: Supporting curriculum designers in developing balanced outcome-based programs using knowledge graphs. In: International Journal of Advanced Trends in Computer Science and Engineering (2023). https://api.semanticscholar.org/CorpusID:264088404

Online Inductive Learning from Answer Sets for Efficient Reinforcement Learning Exploration

Celeste Veronese[(✉)], Daniele Meli, and Alessandro Farinelli

Department of Computer Science, University of Verona, Verona, Italy
{celeste.veronese,daniele.meli,alessandro.farinelli}@univr.it

Abstract. This paper presents a novel approach combining inductive logic programming with reinforcement learning to improve training performance and explainability. We exploit inductive learning of answer set programs from noisy examples to learn a set of logical rules representing an explainable approximation of the agent's policy at each batch of experience. We then perform answer set reasoning on the learned rules to guide the exploration of the learning agent at the next batch, without requiring inefficient reward shaping and preserving optimality with soft bias. The entire procedure is conducted during the online execution of the reinforcement learning algorithm. We preliminarily validate the efficacy of our approach by integrating it into the Q-learning algorithm for the Pac-Man scenario in two maps of increasing complexity. Our methodology produces a significant boost in the discounted return achieved by the agent, even in the first batches of training. Moreover, inductive learning does not compromise the computational time required by Q-learning and learned rules quickly converge to an explanation of the agent's policy.

Keywords: Inductive Logic Programming · Neurosymbolic AI · Reinforcement Learning · Explainable AI · Answer Set Programming

1 Introduction

The intersection of symbolic, logic-based reasoning and Reinforcement Learning (RL) represents a trending topic in Artificial Intelligence (AI) research, aiming to harness the strengths of both domains and address their individual limitations [19]. Current state-of-the-art approaches in model-free RL, in fact, predominantly rely on extensive data or pre-defined environmental models, facing significant challenges in terms of efficiency and scalability. Furthermore, the decision process underlying policy generation is not transparent to other agents and humans, hindering certifiability and trustworthiness. On the other hand, symbolic AI works well in the small data regime but does not perform well on non-symbolic data and is not noise tolerant [1]. Incorporating symbolic and logical formalisms into RL systems, as highlighted by [20], can significantly boost their interpretability, thereby fostering wider acceptance and diffusion while also playing a crucial role in improving the policy and the training phase. However, symbolic learning and reasoning may significantly increase the computational burden of RL, thus limiting the effective real-time integration of neurosymbolic approaches [8] or the scalability to non-trivial tasks, e.g., where multiple predicates and scopes for logical variables shall be defined [7].

Inspired by results obtained by [4,13] in model-based RL, we present a novel approach that exploits Inductive Logic Programming (ILP, [14]) to iteratively learn human-interpretable policy heuristics *online*, during the training of RL agents, and directly use them to bias the agent in its exploration process. Specifically, for each batch of experiences (state-action sequences) collected by the RL agent during training, we rank them by the cumulative return and convert them to a logical formalism to build examples for ILP. This representation maps states and actions to basic human-level concepts about the task, thus enhancing the interpretability of the trained policy. ILP then learns a logical approximation of the policy, which is then used in the following batch to bias the agent's exploration to improve the performance by avoiding wrong or dangerous actions. We adopt the logical formalism of Answer Set Programming (ASP) [15] to represent logical specifications and perform online heuristic reasoning. ASP is chosen as it can be considered the state-of-the-art in planning domain representation for autonomous agents [6]. In this way, we can rely on the latest developments in scalable and efficient ILP, even in the presence of many examples [16]. We preliminarily apply our methodology on the simple approximate Q-learning algorithm [5], to highlight better the variations introduced by our approach. However, any RL algorithm that includes a random exploration phase could be used (e.g. Double Deep Q-Network [3]). In summary, this paper presents the following contributions to the state-of-the-art:

- we introduce an innovative approach to online neurosymbolic learning and reasoning. Our methodology does not simply shape the reward of the agent according to the learned logical heuristics, but directly guides RL exploration, resulting in a tighter and more efficient neurosymbolic interconnection [19]. Moreover, we preserve RL optimality guarantees adopting a probabilistic soft bias approach;
- differently from [13], we perform the learning of informative logical heuristics *online* during RL training while maintaining the expressiveness of ASP. To this aim, we exploit fast scalable ILP as in [16];
- we empirically evaluate the effectiveness of our methodology in the benchmark Pac-Man scenario on maps of increasing complexity with contrastive agents. We obtain a significant boost in the discounted return achieved by the agent. Furthermore, while the inductive learning component minimally affects computational time, it facilitates a swift convergence of learned rules, leading to a coherent and comprehensive explanation of the underlying black-box policy.

2 Related Work

Recent works demonstrate that incorporating existing knowledge into RL and Markov Decision Processes (MDPs) can greatly enhance the development of effective policies [19]. The integration of such knowledge through logical formalisms has significantly improved policy computation. For instance, the REBA framework by [21] employs ASP to define spatial relations in a domestic setting, guiding a robotic agent to select specific rooms for inspection while addressing more straightforward MDP tasks locally. Similarly, DARLING [27] uses ASP to restrict MDP exploration in simulated grid environments and real-world robotics applications. Furthermore, [22,23] utilize linear temporal logic to direct exploration in MDPs. Logical constraints also play a crucial role in avoiding undesirable behaviours, particularly in safety-sensitive scenarios [24,25]. However, since logical heuristics usually express policy-related knowledge (unknown to the agent), they need to be defined by expert users. In the alternative, a huge amount of high-quality training examples are needed to learn them, such as in [26], in which temporal logic specifications are learnt as finite automata from good example traces and used to shape the reward signal. For this reason, attempts have been made to combine neural and symbolic learning. For instance, in [7] Neural Markov Logic Networks are used to learn relational representations from structured examples. However, the approach is computationally inefficient out of very simple domains, involving few variables and predicates. In [28] a deep relational reinforcement learning approach to learn generalizable policies is proposed. Nonetheless, the agent only adopts the logical policy at training convergence, thus not fully exploiting the synergy with neural methods, which are inherently more noise-robust and can help refine the agent's performance. Furthermore, even the solution presented in [28] suffers from poor scalability, requiring task-specific constraints on the available search space for logical policies.

In this paper, we combine deep RL with an ILP framework under the ASP semantics, which offers great expressiveness for structured task representation and reasoning in a fragment of first-order logic [6]. ILP under the ASP semantics has been proven successful in numerous scenarios, e.g., in enhancing the explainability of black-box models [29,30] and gaining task knowledge [9,31]. Among different available implementations [11,12], we adopt the popular Fast-LAS system [16], which can scale to very large search spaces. In this way, we can efficiently learn ASP policy approximations online while training the RL agent, solving the computational limitations posed by [7,28].

Our work is close to the one in [8], where ASP rules are learned to define reward machines in RL settings. However, our solution is more scalable thanks to the FastLAS approach. Furthermore, we do not use rules to shape the reward signal of the RL agent, which may be inefficient in effectively boosting the performance of RL training since still many interactions with the environment may be required to learn the value of action [19]. Instead, we take inspiration from [13] and perform ASP reasoning over the learned policy heuristics to bias the exploration of an RL agent with a higher probability towards heuristic-suggested actions.

3 Background

We now introduce some basic notions about the approximate Q-Learning algorithm, ASP and ILP, and describe the benchmark domain we used to test our methodology.

3.1 Approximate Q-Learning

Q-learning is a model-free reinforcement learning algorithm used to find an optimal action-selection policy [5]. The goal of the agent is to learn the optimal policy that maximizes the expected cumulative reward over time. The core concept in Q-learning is the action-value function, $Q(s,a)$, which estimates the expected future rewards for taking action a in state s. The Q-learning update rule is given by:

$$Q(s_t, a_t) \leftarrow Q(s_t, a_t) + \alpha \left(r_{t+1} + \gamma \max_{a'} Q(s_{t+1}, a') - Q(s_t, a_t) \right)$$

where s_t is the current state, a_t is the action taken in that state, r_{t+1} is the reward received after taking action a_t, and s_{t+1} is the next state. The parameter $\alpha \in [0,1]$ is the learning rate, which controls how much new information overrides old information, and $\gamma \in [0,1]$ is the discount factor, which determines the importance of future rewards. The algorithm converges to the optimal action-value function $Q^*(s,a)$ under certain conditions, such as a decaying learning rate and sufficient exploration of all state-action pairs. The optimal policy $\pi^*(s)$ is then derived by choosing the action that maximizes the Q-value in each state:

$$\pi^*(s) = \arg\max_{a} Q^*(s,a).$$

Approximate Q-learning addresses scalability issues inherent in traditional Q-learning, particularly in environments with large or continuous state spaces [5]. The key idea is to approximate the Q-function using a combination of features instead of the full state. That is, instead of recording everything in detail, we think about what is most important to know, and model that.

3.2 Answer Set Programming

Answer Set Programming (ASP) defines a domain as a set of logical statements (axioms) articulating the logical relationships between entities represented as predicates and variables (atoms) [18]. Axioms considered in this work are *normal rules* $h \text{ :- } b_1, \ldots, b_n$, which define the body of the rule (i.e. the logical conjunction of literals $\bigwedge_{i=1}^{n} b_i$) as a precondition for the head h. We say that a variable is grounded when assigned a particular value. Consequently, an atom is grounded when all its variables are grounded. Given an ASP program P, its Herbrand base $\mathcal{H}(P)$ defines the set of ground atoms which can be generated from it. From an ASP domain definition, an ASP solver computes the *answer sets*, i.e., the minimal set of literals that satisfies the given logic program according to the stable model semantics. This work assumes that body atoms represent environmental features describing S in the MDP, while head atoms are actions. Answer sets will then contain feasible actions for the RL agent.

3.3 Inductive Logic Programming

A generic ILP task under a logical formalism F [14] is defined as a tuple $\mathcal{T} = \langle B, S_M, E \rangle$, consisting of background knowledge B expressed in a logic formalism F, a search space S_M containing the set of possible axioms to be learned (defined, e.g., via a mode declaration M [33]), and a set of examples E, all expressed in the syntax of F. The goal is to find a hypothesis (i.e. an axiom) $H \subseteq S_M$ covering E. Under the ASP semantics, we consider the generic case where examples are *weighted context dependent partial interpretations* (WCDPI's) [32]. A partial interpretation e^{pi} is a pair of sets of ground atoms $\langle e^{inc}, e^{exc} \rangle$. An interpretation (i.e., a set of ground atoms) I extends e iff $e^{inc} \subseteq I$ and $e^{exc} \cap I = \emptyset$. A WCDPI is then a tuple $e = \langle e^{id}, e^{pen}, e^{pi}, e^{ctx} \rangle$, where e^{id} is an identifier for e, e^{pen} is either a positive integer or ∞, called a penalty, e^{pi} is a partial interpretation, and e^{ctx} is an ASP program called the context. A WCDPI e is accepted by a program P iff there is an answer set of $P \cup e^{ctx}$ that extends e^{pi}. Following the definition by [32], the goal of ILP is then to find a hypothesis $H \subseteq S_M$ with minimal length (i.e., number of atoms) and $\sum_e e^{pen}$, for all examples e not accepted by $H \cup B$. Since we are interested in discovering normal rules matching actions to environmental features, e^{inc}, e^{exc} represent executed and not executed actions, respectively, while e^{ctx} is the set of ground environmental features. We employ FastLAS learner by [16] for fast computation from acquired examples gathered from RL batches of experience.

3.4 The Pac-Man Domain

In the Pac-Man domain, an agent (Pac-Man) needs to navigate a maze-like environment to collect food pellets (each one gives a +10 reward) while avoiding enemies (ghosts). The G ghosts present in the environment generally move randomly but may start chasing Pac-Man (with probability p_g) if they are close (we conducted our experiments with $p_g = 0.8$). Given the $N \times M$ grid, the agent can move in the four cardinal directions or stay still (hence $|A| = 5$), receiving a time penalty of -1 for each time step spent in the maze. The episode ends with a +500 reward if Pac-Man eats all food pellets or a -500 penalty if one of the ghosts chases Pac-Man. The environment also contains P power capsules (big yellow dots in Fig. 1) that Pac-Man can eat to gain the ability to scare and eat ghosts. Catching a scared ghost results in a +200 reward. The state S contains the position of each wall, food pellet, power pill, ghost in the map, and the Pac-Man position; each is expressed as (x, y) coordinates. The environment is fully deterministic. The challenge Pac-Man presents derives from the vast dimensions of the state space and the extended planning horizon: to complete the level, the agent may have to explore the entire environment to find all the pellets, and non-trivial movements are often required to escape ghosts.

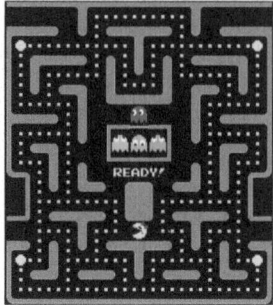

Fig. 1. Example scenario for the Pac-Man domain, $G = P = 4$, 25×26 grid.

4 Methodology

This section describes our methodology for integrating approximate Q-learning, symbolic learning, and reasoning in ASP. We exemplify it in the context of the standard RL Pac-Man domain used for empirical evaluation.

4.1 ASP Representation of the Domain

We start from the representation of the domain in ASP syntax. This requires defining environmental features $\mathcal{F}$ and actions $\mathcal{A}$. For the Pac-Man domain, $\mathcal{F}$ contains: wall(Dir), which denotes the presence of a wall in front of the agent (i.e., one cell away) in that direction, food(Dir, Dist), ghost(Dir, Dist), and capsule(Dir, Dist), representing Manhattan distance Dist $\in [0, 10]$ between PacMan and a cell containing food (or ghost, or capsule) in direction Dir $\in$ {north, south, east, west}. Finally, we need to introduce upper and lower bounds on Dist to obtain more informative policy heuristics. To this aim, we define atoms in the form X_dist_Y(Dir,Dist,D), where X is either ghost, food or caps, and Y is either geq or leq, defined as follows:

X_dist_geq(Dir, Dist, D) :- X(Dir, Dist), Dist >= D, d_const(D).
X_dist_leq(Dir, Dist, D) :- X(Dir, Dist), Dist <= D, d_const(D).

where d_const(0..4) limits the set of possible values that D can take in the rule. Action atoms are constructed from A as move(Dir), denoting the movement of the agent in direction Dir. The agent also has the option to perform a 'stop' action. However, we chose not to include it in the learning phase as it was rarely performed in the examples collected during our tests. Once $\mathcal{F}$ and $\mathcal{A}$ are defined, we need a *feature map* $F_\mathcal{F} : S \to \mathcal{H}(\mathcal{F})$ and an *action map* $F_\mathcal{A} : A \to \mathcal{H}(\mathcal{A})$ to ground atoms from collected batches of RL. The only information needed to build $\mathcal{F}$ is the positions of the agent, food, ghosts and capsules in the environment, all of which are available in S.

4.2 Definition of the Learning Task

Given the ASP formalization of the task, we need to generate the ILP task $T = \langle B, S_M, E \rangle$ for FastLAS, starting from RL episodes. Given an episode consisting of a sequence of N state-action pairs $\langle \bar{a}_i, \bar{s}_i \rangle$, $i = 1, \ldots N$, we can build a WCDPI of the following form:

$$e_i = \langle id_i, w_i, \langle \{\bar{a}_i\}, a \in F_{\mathcal{A}} \setminus F_{\mathcal{A}}(\bar{a}_i) \rangle, F_{\mathcal{F}}(\bar{s}_i) \rangle,$$

where id_i is a unique identifier and w_i represents the penalty of the WCDPI, set as the reward obtained by the agent in that episode and assigning $w_i = 0$ to those episodes that obtained a negative reward.[1] This gives more relevance to the agent's behaviour, resulting in a higher reward for generating rules that can lead to at least the same performance. We populate e^{inc} with the agent's chosen action and e^{exc} with the grounding for all unobserved actions. For example, considering the Pac-Man scenario depicted in Fig. 1, if at time t the agent moves left within an episode with return 50, we generate the following WCDPI:

$$e_t = \langle id_t, 50, \langle \text{move(east)}, \{\text{move(west)}, \text{move(north)}, \text{move(south)}\} \rangle,$$
$$\{\text{wall(north)}, \text{wall(south)}, \text{food(east, 1)}, \text{food(west, 1)}, \ldots\} \rangle, \quad (1)$$

We have omitted ground context atoms with a distance higher than 1 for simplicity. The background knowledge of the task only contains the definition of the ASP variables and ranges, and the mode declaration is defined in order only to have rule's heads in the form move(Dir), and bodies containing atoms from $\mathcal{F}$.

4.3 Neurosymbolic Q-Learning

Algorithm 2 shows our Neurosymbolic Q-Learning methodology. At the end of each batch of experience acquired by Approximate Q-Learning, we store at most σ highest-return episodes (Line 22). We then follow the procedure described in Sect. 4.2 to generate WCDPIs and learn policy heuristics H (Line 23). These heuristics, together with the background knowledge B defined in Sect. 4.2 and the grounding of state variables $F_{\mathcal{F}}(s)$, are used to perform ASP reasoning and compute a set of *suggested actions* A_h at each step of the Q-learner (Line 7). Specifically, in the exploration phase of the agent (with probability ϵ), it may choose either an action from A_h or $A \setminus A_h$ (i.e., actions which H does not suggest), with probability ρ and $1 - \rho$ respectively (Line 10). In this way, we preserve the asymptotic optimality of the Q-learning algorithm [5], following a soft bias approach as in existing literature [13]. We empirically choose ρ as the average discounted return over the best-saved episodes E_σ, normalized by the average return of the whole training. In order not to start with $A_h = \emptyset$ until

---

[1] This is necessary because FastLAS does not handle negative weights. However, as explained in the following section, we only consider the best episodes of each batch; thus, the presence of rewards less than zero is highly improbable.

the first batch of episodes is gathered, we use the very first episode of training to generate preliminary policy heuristics (Line 21), which provides a greedy yet useful hint to the agent, as explained in Sect. 5. We remark that the described methodology can be equivalently applied to any RL algorithm, where random actions are taken in the exploration phase [5].

Algorithm 2 Neurosymbolic Q-Learning

Require: MDP $= \langle S, A, R, T, \gamma \rangle$, background knowledge B, search space S_M
Parameters: Q-learning parameters [5] (including $\epsilon \in [0,1]$), max best episodes stored σ, max episodes E, batch size S_b

1: Initialize weight vector $\mathbf{w}$, batch $b = \emptyset$, episode $e = \emptyset$, episode count $N_e = 0$, policy heuristics $H = \emptyset$, set of best episodes $E_\sigma = \emptyset$
2: **while** $N_e < E$ **do**
3:    Observe initial state s
4:    **while** s is not terminal **do**
5:       $x \sim [0,1]$
6:       **if** $x < \epsilon$ **then**
7:          $A_h \leftarrow \mathbf{ASP}(B, H, F_\mathcal{F}(s))$
8:          **if** $A_h \neq \emptyset$ **then**
9:             $\rho \leftarrow \mathbf{AvgReturn}(E_\sigma)$
10:           $a \sim \mathbf{WeightedProb}(A, F_\mathcal{A}^{-1}(A_h), \rho)$
11:          **else**
12:             $a \sim A$
13:          **end if**
14:       **else**
15:          $a \leftarrow \arg\max_{a'} Q(\mathbf{w}, s, a')$
16:       **end if**
17:       $e.\mathrm{Append}(\langle s, a \rangle)$
18:       $s' \leftarrow T(s,a),\ r \leftarrow R(s,a,s')$
19:       $\mathbf{w} \leftarrow \mathbf{QUpdate}(s, a, r, s', \gamma)$
20:       $s \leftarrow s'$
21:    **end while**
22:    $N_e \leftarrow N_e + 1$
23:    $b.\mathbf{Append}(e)$
24:    **if** $|b| = S_b \vee N_e = 1$ **then**
25:       $E_\sigma \leftarrow \mathbf{Update}(b, \sigma)$
26:       $H \leftarrow \mathbf{FastLAS}(B, S_M, E_\sigma)$ {E_σ converted to ASP syntax via $F_\mathcal{F}, F_\mathcal{A}$}
27:    **end if**
28: **end while**

---

5 Empirical Evaluation

Experimental results on the methodology outlined in the previous section are now presented. We tested our approach on two different scenarios of the Pac-Man domain: a smaller map, with 18×9 grid dimensions, $P = 2$ power capsules and $G = 2$ ghosts, and a way more challenging map, with 25×26 grid dimensions, $G = 4$ ghosts and $P = 4$ power capsules. All experiments were performed on a computer equipped with 5.1GHz, a 13th Gen Intel i5 processor and 64GB RAM. We evaluate 2 different performance measures:

- *training performance*: it measures the RL discounted return and computational efficiency;
- *policy heuristic convergence*: it measures the explainability and robustness of the symbolic component of our methodology, evaluating how the set of policy heuristics reaches a fixed point as RL training progresses.

For all metrics, we compare our methodology (**NeuroQ**) against classical Approximate Q-Learning (**ApproxQ**). We report results as mean and standard deviation over a set of 5 random seeds for statistical relevance. For each seed, we choose a maximum number of episodes $E = 20000$ and a batch size $S_b = 100$ (resulting in 200 training batches). Q-Learning parameters for both evaluated algorithms are set as follows: learning rate $\alpha = 0.2, \gamma = 0.8, \epsilon = 0.05$. We empirically choose to keep track of $\sigma = 5$ best episodes on the smaller map, while on the bigger map (in which longer episodes are generated), we set $\sigma = 3$. In this way, we balance RL performance (discounted return) and the computational cost of symbolic learning.

5.1 Training Performance

Figure 2 shows the return obtained by the execution of ApproxQ and NeuroQ in the small map (2a) and in the more challenging map (2b). We note that, in both scenarios, the introduction of policy heuristics significantly increases the return obtained by the agent, leading to a more efficient training process. In addition to assessing the trade-off between performance and computation, we investigated the computational impact of generating policy heuristics and calculating the suggested actions. The execution times, shown in Tables 1, 2 for the small and large maps, respectively, demonstrate that the additional time required by FastLAS learning and ASP reasoning (Lines 7 and 23 in Algorithm 2) is acceptable, given the substantial increase of performance. In particular, in the large map, NeuroQ requires $\approx 25\%$ more time per batch, but the average discounted return is double (Fig. 2b).

5.2 Policy Heuristics Convergence

At each batch iteration of Algorithm 2, the ILP task solved by FastLAS consists of ≈ 650 WCDPIs (on average) for both the small and large maps. The search space S_M remains constant during the procedure and consists of 226 unique rules for both environments. As the first training episode is acquired (Line 21 of Algorithm 2), we learn a first set of policy heuristics to bias the agent's

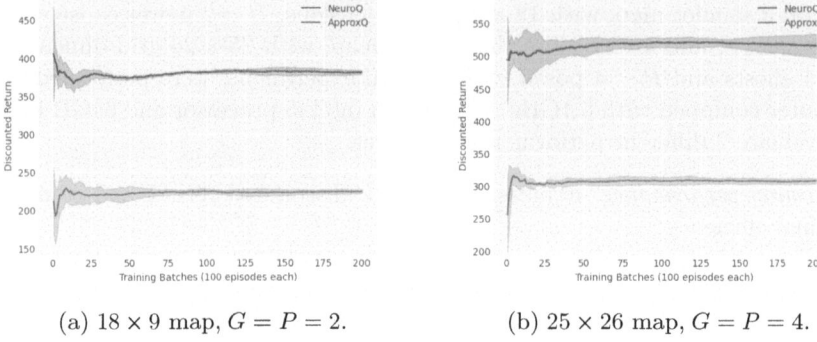

(a) 18 × 9 map, $G = P = 2$. (b) 25 × 26 map, $G = P = 4$.

Fig. 2. Average discounted return in the Pac-Man domain.

Table 1. Execution times (in seconds) in total and per batch, on the 18 × 9 map with $G = P = 2$. Mean and standard deviation (in brackets) are reported where appropriate.

Seed	ApproxQ		NeuroQ	
	Total	Per batch	Total	Per batch
0	949.39	4.74 (±0.67)	1505.91	7.53 (±0.57)
1	955.86	4.78 (±0.65)	1497.01	7.48 (±0.60)
2	953.72	4.77 (±0.67)	1508.62	7.54 (±0.56)
3	952.75	4.76 (±0.68)	1496.73	7.48 (±0.57)
4	953.46	4.77 (±0.67)	1503.07	7.52 (±0.60)
Average	953.04 (±2.34)	4.77 (±0.67)	1502.27 (±4.92)	7.51 (±0.58)

exploration towards more convenient actions immediately. On the 18 × 9 map, for example, the following heuristic is generated:

$$\texttt{move(Dir) :- food\_dist\_leq(Dir,Dist,1).}$$

In other words, the agent immediately learns to move in the direction where the food is close. In the 25 × 26 map, FastLAS learns a similar heuristic at the first iteration, with slight adjustments depending on the specific seed (e.g., not wall(Dir) is included in the body, too).

As the training progresses, learned heuristics finally converge to the following in the smaller environment:

$$\texttt{move(Dir) :- food\_dist\_leq(Dir,Dist,1).} \quad (2a)$$
$$\texttt{move(Dir) :- caps\_dist\_leq(Dir,Dist,1).} \quad (2b)$$

The first greedy heuristic learned in the first episode is confirmed, while the second rule pushes the agent towards close power capsules. On the 25 × 26 map,

Table 2. Execution times (in seconds) in total and per batch, on the 25 × 26 map with $G = P = 4$. Mean and standard deviation (in brackets) are reported where appropriate.

Seed	ApproxQ		NeuroQ	
	Total	Per batch	Total	Per batch
0	3980.74	19.9 (±3.26)	4974.46	24.8 (±3.44)
1	4005.92	20.0 (±3.46)	5233.66	26.1 (±3.20)
2	4054.08	20.2 (±3.37)	5197.47	26.0 (±3.16)
3	4003.02	20.0 (±3.45)	5196.34	26.0 (±3.26)
4	4025.39	20.1 (±3.13)	5174.59	25.9 (±3.35)
Average	4013.83 (±27.51)	20.07 (±3.33)	5155.30 (±105.7)	25.78 (±3.28)

the final set of learned rules is the following:

move(Dir) :- caps_dist_leq(Dir,Dist,1).   (3a)
move(Dir) :- food_dist_leq(Dir,Dist,1).   (3b)
move(Dir) :- not wall(Dir), food_dist_leq(Dir,Dist,2).   (3c)
move(Dir) :- ghost_dist_geq(Dir,Dist,4).   (3d)

In this case, the agent interacts with a more complex map for a longer time. As a consequence, the policy heuristics are more informative and also capture the necessity of avoiding ghosts and walls, stating that the agent should move in a direction only if no walls are present (not wall(Dir)) or if ghosts are at a sufficiently high distance (ghost_dist_geq(Dir,Dist,4)) in that direction.

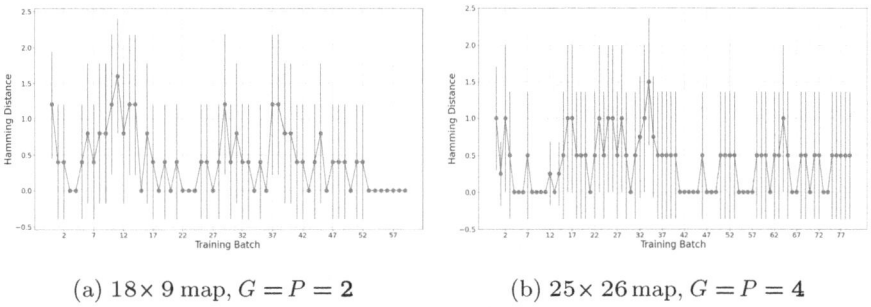

(a) 18× 9 map, $G = P = 2$ (b) 25× 26 map, $G = P = 4$

Fig. 3. Convergence of learned policy heuristics (mean ± standard deviation over seeds). To facilitate visualization, batches beyond convergence are omitted.

Figure 3 shows the convergence of the hypothesis for the small (3a) and large (3b) maps, respectively. Convergence per batch is measured as the Hamming distance between the literals composing the body of rules extracted at a given batch, normalized by the number of literals composing the final sets of rules.

Ideally, we aim for zeroing Hamming distance when convergence is achieved. The charts show that the symbolic learner converges within < 70 batches (over 200) in both scenarios, reflecting the convergence of the overall RL policy. In particular, on the large map (Fig. 3b) rules don't converge to a zeroed Hamming distance (averaging over 5 seeds). Indeed, with one seed, FastLAS alternatively learns two sets of rules, one in which rule 3b is present and one in which it is not. However, rule 3c is always present and is more informative than 3b. Hence, this doesn't represent a significant change in the semantics of the policy.

6 Conclusion and Future Work

The research presented in this paper demonstrates how to effectively integrate neurosymbolic learning and reasoning to improve the efficiency and explainability of RL agents. We leverage the expressive ASP semantics to represent structured task knowledge in a fragment of first-order logic and reason over the most convenient actions in the RL exploration phase. ASP policy heuristics are learned and refined online via a scalable ILP algorithm, gathering examples at each batch of RL training. Moreover, a probabilistic soft bias approach in ASP guidance preserves the convergence guarantees of RL. We validated our algorithm in the Pac-Man benchmark RL scenario in two maps of increasing complexity involving long planning horizons, large state space and contrastive agents (ghosts). Our neurosymbolic RL algorithm extends Q-Learning, but it could be applied to any RL algorithm involving a random exploration phase. Our methodology achieves a significantly higher discounted return than traditional RL (almost double in the largest map), at the cost of $\approx 25\%$ increase in the computational cost. In addition, learned ASP rules converge within $\approx< 70/200$ RL batches, providing an interpretable logical explanation of the black-box RL policy and training process. As a future work, we plan to test the scalability of these techniques to even more complex and varied environments, also investigating the integration with other RL algorithms and more complex forms of neurosymbolic integration [19].

References

1. Vermeulen, A., Manhaeve, R., Marra, G.: An experimental overview of neural-symbolic systems. In: International Conference on Inductive Logic Programming (2023). https://api.semanticscholar.org/CorpusID:266965515
2. Mnih, V., Kavukcuoglu, K., Silver, D., et al.: Human-level control through deep reinforcement learning. Nature **518**, 529–533 (2015). https://doi.org/10.1038/nature14236
3. Hasselt, H.V., Guez, A., Silver, D.: Deep reinforcement learning with double q-learning. In: AAAI Conference on Artificial Intelligence (2015). https://api.semanticscholar.org/CorpusID:6208256
4. Mazzi, G., Meli, D., Castellini, A., Farinelli, A.: Learning logic specifications for soft policy guidance in POMCP. In: Proceedings of the 2023 International Conference on Autonomous Agents and Multiagent Systems, pp. 373–381 (2023)

5. Sutton, R.S., Barto, A.G.: Reinforcement Learning: An Introduction, 2nd ed., MIT Press (2018)
6. Meli, D., Nakawala, H., Fiorini, P.: Logic programming for deliberative robotic task planning. Artif. Intell. Rev. **56**(9), 9011–9049 (2023)
7. Marra, G., Kuželka, O.: Neural markov logic networks. In: Uncertainty in Artificial Intelligence, pp. 908–917. PMLR (2021)
8. Furelos-Blanco, D., Law, M., Jonsson, A., Broda, K., Russo, A.: Induction and exploitation of subgoal automata for reinforcement learning. J. Artif. Intell. Res. **70**, 1031–1116 (2021)
9. Meli, D., Sridharan, M., Fiorini, P.: Inductive learning of answer set programs for autonomous surgical task planning. Mach. Learn. **110**(7), 1739–1763 (2021). https://doi.org/10.1007/s10994-021-06013-7
10. Acharya, K., Raza, W., Dourado, C., Velasquez, A., Song, H.H.: Neurosymbolic reinforcement learning and planning: a survey. IEEE Transactions on Artificial Intelligence (2023)
11. Hocquette, C., Muggleton, S.H.: Complete bottom-up predicate invention in metainterpretive learning. In: Proceedings of the Twenty-Ninth International Joint Conferences on Artificial Intelligence, pp. 2312–2318 (2021)
12. Cropper, A., Morel, R.: Learning programs by learning from failures. Mach. Learn. **110**(4), 801–856 (2021). https://doi.org/10.1007/s10994-020-05934-z
13. Meli, D., Castellini, A., Farinelli, A.: Learning logic specifications for policy guidance in POMDPs: an inductive logic programming approach. J. Artif. Intell. Res. **79**, 725–776 (2024)
14. Muggleton, S.: Inductive logic programming. N. Gener. Comput. **8**(4), 295–318 (1991)
15. Lifschitz, V.: Answer set planning. In: International Conference on Logic Programming and Nonmonotonic Reasoning, pp. 373–374. Springer (1999)
16. Law, M., Russo, A., Bertino, E., Broda, K.: FastLAS: scalable inductive logic programming incorporating domain-specific optimisation criteria. Proc. AAAI Conf. Artif. Intell. **34**(3), 2877–2885 (2020). https://doi.org/10.1609/aaai.v34i03.5678
17. Law, M., Russo, A., Broda, K., Bertino, E.: Scalable non-observational predicate learning in ASP. In: Proceedings of the Thirtieth International Joint Conference on Artificial Intelligence (IJCAI-21) (2021)
18. Gelfond, M., Lifschitz, V.: The stable model semantics for logic programming. In: ICLP/SLP (1988). https://api.semanticscholar.org/CorpusID:261517573
19. Cheng, C.-A., Kolobov, A., Swaminathan, A.: Heuristic-guided reinforcement learning. In: Advances in Neural Information Processing Systems, vol. 34, pp. 13550–13563 (2021)
20. Kambhampati, S., Sreedharan, S., Verma, M., Zha, Y., Guan, L.: Symbols as a lingua franca for bridging human-AI chasm for explainable and advisable AI systems. Proc. AAAI Conf. Artif. Intell. **36**, 12262–12267 (2022)
21. Sridharan, M., Gelfond, M., Zhang, S., Wyatt, J.: REBA: a refinement-based architecture for knowledge representation and reasoning in robotics. J. Artif. Intell. Res. **65**, 87–180 (2019)
22. De Giacomo, G., Iocchi, L., Favorito, M., Patrizi, F.: Foundations for restraining bolts: reinforcement learning with LTLf/LDLf restraining specifications. Proc. Int. Conf. Autom. Plann. Sched. **29**, 128–136 (2019)
23. Leonetti, M., Iocchi, L., Patrizi, F.: Automatic generation and learning of finite-state controllers. In: International Conference on Artificial Intelligence: Methodology, Systems, and Applications, pp. 135–144 (2012)

24. Marzari, L., Marchesini, E., Farinelli, A.: Online safety property collection and refinement for safe deep reinforcement learning in mapless navigation. In: 2023 IEEE International Conference on Robotics and Automation (ICRA), pp. 7133–7139 (2023). https://doi.org/10.1109/ICRA48891.2023.10161312
25. Mazzi, G., Castellini, A., Farinelli, A.: Risk-aware shielding of partially observable monte Carlo planning policies. Artif. Intell. **324**, 103987 (2023)
26. De Giacomo, G., Favorito, M., Iocchi, L., Patrizi, F.: Imitation learning over heterogeneous agents with restraining bolts. Proc. Int. Conf. Autom. Plann. Sched. **30**, 517–521 (2020)
27. Leonetti, M., Iocchi, L., Stone, P.: A synthesis of automated planning and reinforcement learning for efficient, robust decision-making. Artif. Intell. **241**, 103–130 (2016)
28. Hazra, R., De Raedt, L.: Deep explainable relational reinforcement learning: a neuro-symbolic approach. In: Joint European Conference on Machine Learning and Knowledge Discovery in Databases, pp. 213–229. Springer (2023)
29. Veronese, C., Meli, D., Bistaffa, F., Rodríguez-Sot, M., Farinelli, A., Rodríguez-Aguilar, J.A.: Inductive Logic Programming for transparent alignment with multiple moral values. CEUR Workshop Proc. **3615**, 84–88 (2023)
30. Rabold, J., Siebers, M., Schmid, U.: Explaining black-box classifiers with ILP–empowering LIME with Aleph to approximate non-linear decisions with relational rules. In: International Conference on Inductive Logic Programming, pp. 105–117. Springer (2018)
31. Rodriguez, I.D., Bonet, B., Romero, J., Geffner, H.: Learning first-order representations for planning from black box states: new results. Proc. Int. Conf. Principles Knowl. Represent. Reasoning **18**, 539–548 (2021)
32. Law, M., Russo, A., Broda, K.: Inductive learning of answer set programs from noisy examples. In: Advances in Cognitive Systems (2018)
33. Muggleton, S.: Inverse entailment and Progol. New Generation Computing, vol. 13, pp. 245–286. Springer (1995)

Author Index

A
Adrian, Weronika T. 82
Aishwaryaprajna 35
Amigoni, Francesco 69
Azzalini, Davide 69

B
Borroto, Manuel 1

C
Castelletti, Andrea 69
Cerutti, Federico 69
Costa, Vítor Santos 52

D
Dong, Hang 35
Dutra, Inês 52

F
Farinelli, Alessandro 93
Ferrero, Marco Adriano 69

G
Gacek, Paulina 82
Giuliani, Matteo 69

I
Ielo, Antonio 1

M
Mazzotta, Giuseppe 1
Meli, Daniele 93
Mesnage, Cédric 35

R
Reis, Luís Paulo 52
Ricca, Francesco 1
Rocha, Filipe Marinho 52

S
Scalise, Marika 22

T
Tosi, Davide 22

V
Veronese, Celeste 93

W
Wang, Xiaoyang 35